T0140727

The role of microRNAs in controlling protein expression noise

Dissertation

zur Erlangung des akademischen Grades

doctor rerum naturalium (Dr. rer. nat.)

im Fach Biophsyik

eingereicht an der

Lebenswissenschaftlichen Fakultät

der Humboldt-Universität zu Berlin

von

Diplom Biophysiker **Jörn Matthias Schmiedel**

geboren am 16.03.1985, Zell

Präsident der Humoldt-Universität zu Berlin

Prof. Dr. Jan-Hendrik Olbertz

Dekan der Lebenswissenschaftlichen Fakultät

Prof. Dr. Richard Lucius

Gutachter/Innen

1.
2.
3.

Tag der mündlichen Prüfung:

Bibliografische Information der Deutschen Nationalbibliothek

Die Deutsche Nationalbibliothek verzeichnet diese Publikation in der
Deutschen Nationalbibliografie; detaillierte bibliografische Daten sind
im Internet über http://dnb.d-nb.de abrufbar.

ISBN 978-3-8325-4216-0

Logos Verlag Berlin GmbH
Comeniushof, Gubener Str. 47,
10243 Berlin
Tel.: +49 (0)30 42 85 10 90
Fax: +49 (0)30 42 85 10 92
INTERNET: http://www.logos-verlag.de

Contents

1 Noise arising from protein production is minimized for noise-prone genes

In this initial chapter I investigate whether mammalian cells adjust intrinsic noise in protein expression for select genes. I make use of the fact that the balance of contributions from transcriptional and post-transcriptional processes to protein production influence protein expression noise and show that noise-prone genes are produced such as to minimize protein expression noise.

1.1 Introduction

Complex multicellular organisms are conglomerates of cells that all contain virtually the same genetic information. Nonetheless, owing to intra- and inter-cellular regulatory programs these cells attain different cellular states. Genuine cellular states are characterized by specific repertoires of expressed genes that facilitate specialized cellular functions within the organism. A high degree of specialization of cellular states presumably requires precise execution of the defining regulatory programs. In order to attain such precision, organisms have to shield themselves against sources of variability. There are three main sources of variability: genetic, environmental, and stochastic.

Genetic variability can arise from mutations or genomic alterations in cells of the living organism. Cells deploy intricate mechanisms to prevent or repair such damage, and ultimately have mechanisms in place that initiate cell destruction (apoptosis) when too much genetic damage has accumulated. The consequences of failures of these mechanisms become apparent in cancer.

Environmental variability is ever present and hardly under the control of organisms. However, multicellular organisms invest considerable resources to shield most of their cells from the outside world and thereby create a constant environment.

Stochastic variability, which I focus on in this thesis, arises from the randomness of chemical reactions. This randomness becomes apparent when reactions are infrequent or involve low number of molecules. In the process of gene expression this is the case for the reactions involving messenger RNAs (mRNAs). Often, infrequent transcriptional events (~1 per hour) and few mRNA molecules (1 to 10) are sufficient to produce thousands of proteins, because translation acts as an amplifying step (Schwanhäusser et al., 2011). Infrequent transcriptional events can induce large relative fluctuations in the number of mRNAs (Raj et al., 2006); and this variability - or

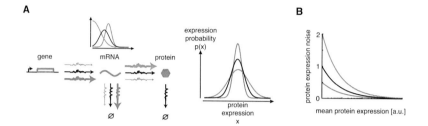

Figure 1: **The balance between transcription and translation influences protein expression noise**

(**A**) Scheme of gene expression with varying transcriptional and post-transcriptional rates and their influence on protein expression noise (the relative width of the distribution, defined as standard deviation over mean). A gene is transcribed into mRNA. mRNA is translated into protein. mRNA and protein are degraded eventually. An intermediate set of rate parameters is indicated in black (with the width of arrows indicating size of particular parameters). A set of parameters with low transcription rate, slow mRNA decay but efficient translation is indicated in orange. A set of parameters with high transcription rate, fast mRNA decay but inefficient translation is indicated in blue. Plots above and to the right exemplify resulting mRNA and protein expression distributions.

(**B**) Protein expression noise as a function of mean protein expression. Noise - the standard deviation divided by the mean of a distribution - decreases with increasing mean expression, due to more frequent transcriptional events and higher mRNA copy numbers. Relative changes in noise due to shifted balances between transcription and translation are independent of the mean expression, and absolute differences are thus larger at low mean expression (colors as in (**A**)).

noise - in the number of mRNA molecules will propagate via translation to the protein level (Figure 1A, black) (Thattai and van Oudenaarden, 2001, Ozbudak et al., 2002, Elowitz et al., 2002, Blake et al., 2003, Paulsson, 2004). In general, noise - commonly quantified as the relative width of a distribution (i.e. standard deviation divided by mean) - in protein numbers is large for lowly expressed genes, but decreases for more strongly expressed genes, because they deploy more frequent transcriptional events and more mRNAs to produce more proteins (Figure 1B, black)(Ozbudak et al., 2002, Elowitz et al., 2002, Bar-Even et al., 2006, Newman et al., 2006). Additionally, external factors, which influence the expression of a gene, propagate noise to the gene (Elowitz et al., 2002, Swain et al., 2002, Pedraza and van Oudenaarden, 2005). Noise is commonly distinguished in intrinsic noise - resulting from the production process of the gene - and extrinsic noise - noise that propagates from external sources

(Elowitz et al., 2002, Swain et al., 2002). In this first chapter I will exclusively focus on intrinsic noise; extrinsic noise will be of importance in the the later chapters. In the following I give a brief overview of the parameters that influence intrinsic noise.

The field of stochastic variability has thriven on a combination of theoretical and experimental analyses. Mathematical models have been used to elucidated how the kinetic parameters that determine the expression of a protein should influence intrinsic protein expression noise; and subsequently these predictions were systematically validated by experiments. An early theoretical study predicted that intrinsic noise in protein expression is mainly determined by fluctuations in the mRNA level due to low copy number noise (Thattai and van Oudenaarden, 2001). This prediction was validated by experiments showing that noise in protein expression decreases when a gene is more strongly transcribed, leading to higher number of mRNAs, but stays constant when instead the translational efficiency is altered (Ozbudak et al., 2002). Therefore at a given mean protein expression, a protein that is produced by transcribing many mRNAs but only inefficiently translating them to protein will have reduced noise in protein expression (Figure 1, blue). On the contrary, a protein that is produced by transcribing few mRNAs but translating them very efficiently will have increased noise in protein expression (Figure 1, orange).

Furthermore, the lifetimes of mRNA and protein play a crucial role in the propagation of fluctuations from the mRNA to the protein level. If the lifetimes of mRNA and protein are similar, the protein will closely track the fluctuations in the mRNA numbers. If instead the protein is longer-lived - as it is commonly the case - , it will effectively average over the more rapid fluctuations in mRNA number, leading to dampened propagation of noise from the mRNA to the protein level (Figure 3A) (Paulsson, 2004, Pedraza and van Oudenaarden, 2005).

Finally, studies have shown that transcriptional dynamics themselves can influence the noise in protein expression. Genes transition between states of transcriptional activity and inactivity (Ross et al., 1994, Walters et al., 1995). Therefore mRNAs are often transcribed in bursts while the gene is in the transcriptional active state, followed by prolonged periods of transcriptional inactivity . The variability in the number of mRNAs increases the more mRNAs are transcribed per burst and the less frequent bursts occur (Blake et al., 2003, Raser and O'Shea, 2004, Becskei et al., 2005, Raj et al., 2006, Pedraza and Paulsson, 2008).

Taken together, genes in principle have a set of parameter choices available allowing to adjust intrinsic noise in protein expression independently from the mean protein expression. Do multicellular organisms use these possibilities to tune intrinsic noise to their needs?

In the following I will explore whether mammalian cells systematically adjust the production of genes in order to tune noise on the protein level. In particular, I hypothesize that genes more prone to noise - because of low protein expression or short protein lifetimes - as well as genes performing critical cellular functions should be produced such that intrinsic noise in their protein expression is minimized.

1.2 Results

In order to analyze the balance of contributions from transcriptional and post-transcriptional processes to the expression of proteins we introduce two measures. First, we define the contribution of post-transcriptional processes to protein production as the number of proteins produced per mRNA lifetime (i.e. the translation rate constant divided by the mRNA degradation rate constant). Second, we will refer to the balance of transcriptional and post-transcriptional processes as the *production ratio*, which we define as the transcription rate divided by the number of proteins produced per mRNA lifetime. A large production ratio means that protein production is mainly achieved via transcribing many mRNAs and will therefore result in low protein expression noise. A small production ratio indicates that protein production is mainly achieved via efficient translation and will therefore result in high protein expression noise.

To explore the production ratios of genes in mammalian cells we analyzed a published dataset containing the expression rate parameters of three thousand genes in mouse embryonic fibroblast NIH3T3 cells (Schwanhäusser et al., 2011). Here mRNA levels and half-lives were measured by pulse-labeling mRNAs with 4-thiouridine, followed by RNA deep-sequencing. Protein levels and half-lives were measured by pulse-labeling proteins with heavy amino acids, followed by mass-spectrometry measurements. Subsequently, a mathematical model was used to infer transcription rates and translation rate constants from the mRNA and protein levels and half-lives.

We first analyzed production ratios as a function of the translation rates of genes. Strikingly, we find that genes exhibit up to two orders of magnitude differences in contributions from transcriptional and post-transcriptional processes to translate similar amounts of protein (Fig. 2). This should - at a given protein lifetime - lead to similar mean protein expression, but substantially different protein expression noise between those genes. Furthermore, the data show that production ratios are on average larger for lowly translated genes than for medium and highly translated genes.

To exclude the possibility that these differences in production ratios arise from general features of protein expression unrelated to noise, we further examined production ratios as a function of the protein half-lives of genes. The protein half-life determines the response time of protein levels upon induced perturbations (Legewie et al., 2008b). A short half-life makes a protein more sensitive to fluctuations in the number of mRNA molecules (Fig. 3A), because it will trace these fluctuations more promptly (Paulsson, 2004, Pedraza and van Oudenaarden, 2005). Thus, if protein expression noise were controlled, we would expect genes with short-lived proteins to

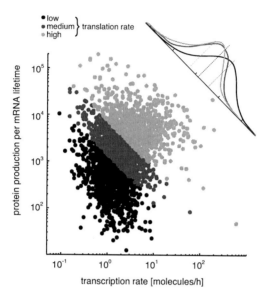

Figure 2: **Genes with similar translation rates are produced with varying contributions from transcriptional and post-transcriptional processes**
Proteins produced per mRNA lifetime as a function of transcription rate for genes in mouse NIH3T3 cells quantified by Schwanhäusser et al. (2011). Genes produced with similar translation rates are indicated by shading. Top right inset shows marginal distribution of production ratios for the three different groups. In the inset solid line indicates median of production ratios for all genes, dashed lines indicate the lower and upper quartiles.

be produced with larger production ratios compared to genes with long-lived proteins. To test this, we grouped genes according to their translation rates and within each group we compared the production ratios of genes with protein half-lives shorter than the median of the group to genes with protein half-lives longer than the median (Fig. 3B, Method Section 4.1). We find that genes with short-lived proteins have larger production ratios, which is consistent with less noise arising in the production of these genes. Reassuringly, differences in production ratios arise from coherent differences in the underlying rates (i.e. higher transcription rates, lower mRNA half-lives and less efficient translation of mRNAs into protein). Again, these differences are more pronounced for genes with low translation rates, indicating that these genes - which are generally more prone to noise - are produced such as to avoid noise resulting from their production.

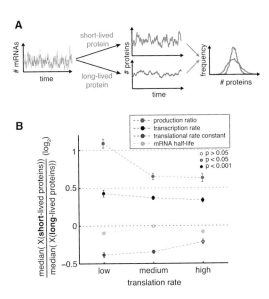

Figure 3: **Short-lived proteins are more noise-sensitive and have larger production ratios compared to long-lived proteins**
(**A**) Protein half-life affects protein expression noise. Stochastic simulations illustrate how fluctuations on mRNA level propagate stronger to short-lived proteins, resulting in higher protein expression noise.
(**B**) Differences in (post-)transcriptional rates between genes with short- and long-lived proteins (shorter or longer than median of bin). Genes were binned together according to their translation rate (bins with equal number of genes, binning as indicated in Figure 2). In each translation rate bin, sets of genes with short- and long-lived proteins were matched for equal translation rate distributions. Error bars represent standard deviation of estimates over 1.000 matched distributions, p-values were computed using two-sided Wilcoxon rank sum test.

However, it is likely that not only expression parameters such as protein half-life determine whether noise in the production of a gene has to be controlled. We expect that the cellular function is another major determinant of whether a gene's protein expression noise has to be controlled for (Fraser et al., 2004). One class of genes for which the cellular function should demand low noise levels are transcription factors. Transcription factors should have low noise levels in order to ensure faithful information transmission and to avoid propagation of noise to downstream genes. We determined 59 transcription factors in the Schwanhäusser et al. (2011) dataset using the Gene Ontology (GOid:3700 "sequence-specific DNA binding transcription factor activity", see Method Section 4.1) (Ashburner et al., 2000). We find transcription factors to be biased towards lower translation rates and shorter protein half-lives. Based on the same binning according to translation rate and protein half-life as before (Figure 3), we asked whether within each bin transcription factors are produced with production ratios larger or smaller than the bin's median. Consistent with our hypothesis that transcription factors should have low noise levels, we find that transcription factors are biased towards larger production ratios, especially those with low translation rates and short and intermediate protein half-lives (Fig. 4). This is again indicative of the notion that noise levels should be more tightly controlled for genes with low expression levels and short-lived proteins, which are more noise-prone.

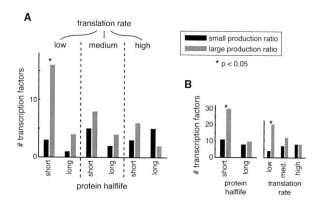

Figure 4: **The expression of transcription factors is biased towards larger production ratios**
(**A**) Number of transcription factors produced with larger (gray) or smaller (black) production ratios compared to median in each bin. Genes were binned according to translation rates then protein half-lives (as in Fig. 3B).
(**B**) To increase statistical power genes were binned only by protein half-life (left) or only by translation rate (right).
(*): $p < 0.05$, Fisher's exact test, multiple testing corrected.

In summary, the results presented here suggest that genes in mammalian cells control intrinsic noise in protein expression noise via their production ratios. This is in line with results from bacteria (Ozbudak et al., 2002) and yeast (Fraser et al., 2004, Lehner, 2008). Genes that are more prone to elevated noise levels due to low protein expression or short protein half-lives as well as genes with noise-sensitive cellular functions, such as transcription factors, have larger production ratios and should therefore have lower intrinsic noise in protein expression than expected.

2 microRNA control of protein expression noise

2.1 Introduction

In the second part of my thesis I investigate the potential of microRNAs to control noise in protein expression. MicroRNAs are the most pervasive post-transcriptional regulators in multicellular organisms and reduce protein expression by means of mRNA destabilization and translational inhibition. Almost all known multicellular organisms possess microRNA genes, some of which have been conserved across million years of evolution (Reinhart et al., 2000, Grimson et al., 2008, Berezikov, 2011). MicroRNAs derive from transcripts that fold into specific stem loop structures (Lee et al., 2002) (Figure 5). These 'primary' microRNAs are then recognized by a set of microRNA biogenesis enzymes that remove the loop and non-pairing overhangs. The resulting double-stranded ~22 nucleotide long RNA is then exported from the nucleus to the cytoplasm. In the cytoplasm one of the two RNA strands is loaded onto Argonaute proteins, which are part of the RNA-induced silencing complex (RISC) (reviewed in Bartel (2004)). The microRNA serves RISC as a guide to complementary regions in mRNAs. At target sites with complete complementary to the microRNA, binding induces catalytic cleavage of the mRNA, similar to small interfering RNAs (Hutvágner and Zamore, 2002). However, target sites with complete complementary to microRNAs are rare and more commonly pairing is established through the microRNA seed region, which encompasses the 5' positions 2 to 7 of the microRNA (Wightman et al., 1993, Lai, 2002, Stark et al., 2003, Lewis et al., 2003). Such pairing is followed by deadenylation of the mRNA's poly(A)-tail, therefore accelerating its decay, and inhibition of its translation (Doench et al., 2003, Lim et al., 2005, Baek et al., 2008, Selbach et al., 2008).

The pairing principles of mRNA-microRNA interactions make it straightforward to identify potential target sites on mRNAs and numerous computational algorithms are available to do so (Krek et al., 2005, Betel et al., 2008, Friedman et al., 2009, Khorshid et al., 2013). However, large fractions of predicted target sites show little endogenous regulatory activity (Lim2005) and algorithms therefore commonly resort to evolutionary conservation of target sites to increase the likelihood of target site functionality (Bartel, 2009). Calculations based on predicted conserved target sites estimated that many microRNAs regulate hundreds of mRNAs each (Enright et al., 2003, Stark et al., 2003, John et al., 2004, Brennecke et al., 2005, Lewis et al., 2005). This results in estimations that the majority of protein-coding genes in higher multicellular organisms contain conserved target sites (Rajewsky, 2006, Friedman

11

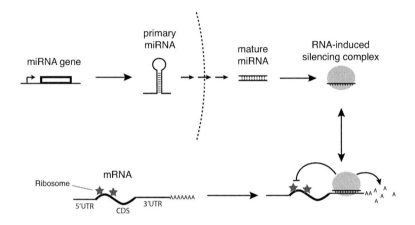

Figure 5: **microRNA biogenesis and regulatory mechanism**
microRNA genes transcribe primary microRNAs that fold into specific stem loop structures. Primary microRNAs are recognized by specific biogenesis enzymes. Biogenesis results in ~22 nucleotide double stranded mature microRNAs that are exported to the cytoplasm. One strand is loaded onto the RNA-induced silencing complex (RISC). MicroRNAs guide RISC to complementary regions on mRNAs, mostly in the 3'UTR. RISC binding results in deadenylation of the poly(A)-tail or translational inhibition.

et al., 2009), often for multiple microRNAs at once (Enright et al., 2003, Stark et al., 2003, Krek et al., 2005).

Predictions on the number of functional targets are in general agreement with experimental studies that perturbed individual microRNAs and observed resulting expression changes of mRNAs (Lim et al., 2005, Linsley et al., 2007, Baek et al., 2008, Selbach et al., 2008, Guo et al., 2010). Surprisingly, these perturbation experiments yielded the conclusion that individual microRNAs repress most of their targets only to a modest degree (less than twofold). Furthermore, studies examining the microRNA knockouts found that such knockouts rarely result in apparent phenotypes (Miska et al., 2007, Alvarez-Saavedra and Horvitz, 2010).

The question therefore arises which significant biological function of microRNAs have that ensures strong conservation of microRNAs and many of their target interactions despite only modest repression and little dependence of organisms on their presence.

The first microRNAs discovered play a role in the timing of *Caenorhabditis el-*

egans larval development and show strong repressive effects on several targets (which is probably the reason why they were discovered by Lee et al. (1993) and Wightman et al. (1993)). These initial and further early discoveries generated a model in which microRNAs were thought of as suppressors of unwanted protein expression, acting as a secondary layer on top of transcriptional repression (Reinhart et al., 2000, Stark et al., 2005). Subsequently, the realization that most mRNA-microRNA interactions only result in modest repression gave rise to an alternative model, proposing that microRNAs rather act to tune protein output (Bartel and Chen, 2004). Furthermore, combining such a tuning function with transcriptional activation and the knowledge about the characteristics of noise resulting from protein production led to the idea that microRNAs might act to reduce protein expression noise (Bartel and Chen, 2004, Hornstein and Shomron, 2006, Herranz and Cohen, 2010, Osella et al., 2011, Ebert and Sharp, 2012).

Based on the explanation given in the first part of this thesis, it will seem plausible to the attentive reader that microRNAs can indeed reduce intrinsic noise in protein production if combined with increased transcriptional activation (Figure 1). However, microRNAs themselves must be subject to fluctuations in their numbers (Figure 6) and such fluctuations should propagate to the regulated gene, resulting in increased extrinsic noise. It is therefore not clear a priori whether microRNA regulation indeed results in reduced overall levels of protein expression noise.

In the following I investigate the effects of microRNAs on intrinsic and extrinsic noise and their relation. In chapter 2.2 is present a detailed derivation of a mathematical model that elucidates the basic properties of microRNA-mediated effects on noise. In chapter 2.3 I present results from single cell reporter experiments that validate the model-based predictions. In chapter 2.4 I attempt to translate these findings to endogenous mRNA-microRNA interactions and investigate the extent of microRNA-mediated noise reduction *in vivo*. Finally chapters 2.5 and 2.6 are concerned with the question of whether microRNA-mediated noise reduction is indeed a significant biological function of microRNAs.

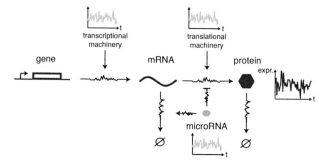

Figure 6: **Noise sources in the expression of a microRNA-regulated gene**
The production of a gene (transcription, translation and decay) results in intrinsic protein expression noise. MicroRNA regulation affects mRNA decay and translation and can therefore alter intrinsic noise. However, microRNAs, like other external regulators, are themselves noise and will therefore add extrinsic noise to the protein expression of the regulated gene.

2.2 Mathematical model of microRNA noise effects

Here I derive expressions for the relative size of fluctuations, or noise, in the protein expression of a microRNA-regulated gene. The calculations are based on a linear noise approximation of the macroscopic description of the respective system, which allows accurate estimates of the extent of fluctuations in the considered molecular species, and how they are correlated between molecular species. A modified version of this chapter will be published as part of Schmiedel et al. (2015).

The macroscopic description of gene expression (Figure 6) used here is based on previously published ordinary differential equation models describing post-transcriptional regulation (Levine et al., 2007a,b, Mehta et al., 2008, Legewie et al., 2008a, Mukherji et al., 2011, Noorbakhsh et al., 2013). In brief, the model describes the transcription and degradation of an mRNA species m; the translation and degradation of the respective protein species p; and assumes that the microRNA μ can reversibly associate with the mRNA to form a mRNA-microRNA complex $m\mu$, which results in accelerated degradation of the mRNA and completely blocks translation of the mRNA into protein.

The ordinary differential equation system is based on the law of mass action and is given by

$$\frac{d[m]}{dt} = \nu^m - d^m \cdot [m] - k^{on} \cdot [m] \cdot [\mu] + \left(k^{off} + d^\mu\right) \cdot [m\mu] \quad, \tag{1}$$

$$\frac{d[m\mu]}{dt} = k^{on} \cdot [m] \cdot [\mu] - k^{off} \cdot [m\mu] - (d^m + d^{m\mu} + d^\mu) \cdot [m\mu] \quad, \tag{2}$$

$$\frac{d[p]}{dt} = k^p \cdot [m] - d^p \cdot [p] \quad, \tag{3}$$

$$\frac{d[\mu]}{dt} = \nu^\mu - d^\mu \cdot [\mu] - k^{on} \cdot [m] \cdot [\mu] + \left(k^{off} + d^m + d^{m\mu}\right) \cdot [m\mu] \quad. \tag{4}$$

Here square brackets denote the concentrations of molecular species. mRNAs are transcribed with the rate ν^m and constitutively degraded with the rate $d^m \cdot [m]$. The reversible association of free mRNAs and free microRNAs is governed by the on-rate $k^{on} \cdot [m] \cdot [\mu]$ and off-rate $k^{off} \cdot [m\mu]$. When bound in the complex, mRNAs are degraded with the additional rate $d^{m\mu} \cdot [m\mu]$, while microRNAs are recycled after the mRNAs are degraded. Proteins are translated from free mRNAs with the rate $k^p \cdot [m]$ and are degraded with the rate $d^p \cdot [p]$. Free microRNA molecules are transcribed at the rate ν^μ and degraded at the rate $d^\mu \cdot [\mu]$. For simplicity we assume that microRNAs are also degraded when bound in the mRNA-microRNA complex, subsequently the mRNAs are then released to their free form.

Adding equations (2) and (4) yields the time evolution of the concentration of total microRNA μ^T

$$\frac{d[\mu^T]}{dt} = \frac{d\left([\mu] + [m\mu]\right)}{dt} = k^\mu - d^\mu \cdot [\mu^T] \quad. \tag{5}$$

It is assumed that association and dissociation of the mRNA-microRNA complex is much faster than the degradation of the mRNA and the microRNA. The quasi-steady-state approximation for the mRNA-microRNA complex thus yields

$$[m\mu] = \frac{[m] \cdot [\mu^T]}{K + [m]} \quad, \tag{6}$$

where $K = \frac{k^{off} + d^m + d^{m\mu} + d^\mu}{k^{on}}$ is the dissociation constant of the mRNA-microRNA complex.

Equation (1) then simplifies to

$$\frac{d[m]}{dt} = \nu^m - d^m \cdot [m] - \frac{(d^m + d^{m\mu}) \cdot [m] \cdot [\mu^T]}{K + [m]} \quad. \tag{7}$$

The steady state concentration of the protein can be expressed as

15

$$[p] = \frac{k^p}{d^p} \cdot \frac{\nu^m}{d^m + \frac{(d^m + d^{m\mu}) \cdot [\mu^T]}{K+[m]}} = \frac{[p_0]}{1 + \frac{d^m + d^{m\mu}}{d^m} \cdot \frac{[\mu^T]}{K+[m]}} \quad , \tag{8}$$

with $[p_0] = \frac{k^p}{d^p} \cdot [m_0]$ as the steady state concentration of proteins (and mRNAs) if the gene were not microRNA regulated.

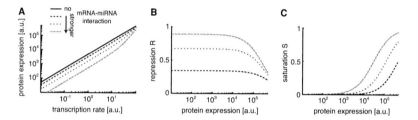

Figure 7: **Effects of microRNA regulation on mean protein expression**
(**A**) Protein expression as function of transcription rate. MicroRNA regulation (gray shades, dashed lines) induces repression compared to a non-microRNA-regulated state (black solid line). Lighter shades of gray indicate stronger mRNA-microRNA interaction (lower dissociation constants K). At high transcription rates, repression declines due to saturation of the microRNA pool (see (C)).
(**B**) Repression strength R increases with increasing mRNA-microRNA interaction strength. For a given mRNA-microRNA interaction strength, repression is maximal and constant at low and intermediate expression levels and declines at high expression levels due to saturation (S).
(**C**) The transcripts of the target gene can saturate the microRNA pool when expressed at high copy numbers. Stronger mRNA-microRNA interaction leads to saturation starting at lower transcript numbers, due to more effective sequestration of the microRNA pool.

Let us examine the influence of microRNA regulation on protein expression. The protein expression of a not microRNA-regulated gene depends linearly on its transcription rate (Figure 7A). MicroRNA regulation of a gene will result in repressed protein expression at equal transcription rates.

To quantify the extent of repression we introduce the measure of repression strength as

$$R = 1 - \frac{[p]}{[p_0]} = \frac{\frac{d^m + d^{m\mu}}{d^m} \cdot \frac{[\mu^T]}{K+[m]}}{1 + \frac{d^m + d^{m\mu}}{d^m} \cdot \frac{[\mu^T]}{K+[m]}} \quad . \tag{9}$$

A repression strength of $R = 0$ thus implies no repression of protein expression and a repression strength of $R = 1$ implies full repression of protein expression. Faster microRNA-induced degradation $d^{m\mu}$, higher total microRNA concentration $[\mu^T]$ as well as a stronger interaction between the mRNA and the microRNA (lower dissociation constants K) will therefore lead to stronger repression (Figure 7A and B).

Of special note is the relation of free mRNA levels to the dissociation constants. If the concentration free mRNA approaches the dissociation constant the repression strength decreases. This can be understood in terms of the limited pool of microRNA molecules. If the concentration of mRNA rises, more microRNA molecules will be bound to mRNAs and the microRNA pool will become saturated. We therefore introduce the measure of saturation as

$$S = \frac{[m]}{K + [m]} = 1 - \frac{[\mu]}{[\mu^T]} \quad , \tag{10}$$

which describes the fraction of total microRNA molecules bound to mRNAs. The concentration of free mRNA equal to the dissociation constant therefore marks the point of half de-repression (Figure 7C).

The relationship between repression strength and total microRNA concentration as well as mRNA-microRNA interaction strength and the effects of saturation have been experimentally verified by Mukherji2011.

Based on the outlined macroscopic description of the system, the linear noise approximation allows to estimate the variances and co-variances in the abundances of the molecular species (Elf and Ehrenberg, 2003). This is accomplished by solving the following fluctuation-dissipation theorem for the entries of the correlation matrix C

$$J \cdot C + C \cdot J^T + D = 0 \quad . \tag{11}$$

Here, the macroscopic rate equations are linearized around the steady state in the Jacobian matrix $J = N \cdot \frac{\partial v}{\partial n}|_{n=\bar{n}}$, with N as the stoichoimetric matrix, v as the vector of fluxes and n as the vector of molecule copy numbers. The diagonal entries in the Jacobian matrix are the time-scales on which fluctuations dissipate in the molecular species (below abbreviated as λ^x), while the off-diagonal entries describe how fluctuations propagate between the molecular species. The diffusion matrix $D = N \cdot diag(v) \cdot N^T$ describes the fluctuation-generating potential of the fluxes through the molecular species.

In brief, the vector of molecule copy numbers is

$$n = \left(\begin{array}{ccc} [m] & [\mu^T] & [p] \end{array} \right) \quad , \tag{12}$$

the vector of fluxes is

$$\nu = \left(\begin{array}{ccccccc} \nu^m & d^m \cdot [m] & \frac{d^{m\mu} \cdot [\mu^T] \cdot [m]}{K+[m]} & \nu^\mu & d^\mu \cdot [\mu^T] & k^p \cdot [m] & d^p \cdot [p] \end{array} \right)^T \quad , \tag{13}$$

and the stoichiometric matrix is

$$N = \begin{pmatrix} 1 & -1 & -1 & 0 & 0 & 0 & 0 \\ 0 & 0 & 0 & 1 & -1 & 0 & 0 \\ 0 & 0 & 0 & 0 & 0 & 1 & -1 \end{pmatrix} \quad . \tag{14}$$

The diffusion matrix is

$$D = \begin{pmatrix} \nu^m + d^m \cdot [m] + \frac{d^{m\mu} \cdot [\mu^T] \cdot [m]}{K+[m]} & 0 & 0 \\ 0 & \nu^\mu + d^\mu \cdot [\mu^T] & 0 \\ 0 & 0 & k^p \cdot [m] + d^p \cdot [p] \end{pmatrix} \quad . \tag{15}$$

The Jacobian matrix is

$$J = \begin{pmatrix} -d^m - \frac{d^{m\mu} \cdot [\mu^T] \cdot K}{(K+[m])^2} & -\frac{d^{m\mu} \cdot [m]}{K+[m]} & 0 \\ 0 & -d^\mu & 0 \\ k^p & 0 & -d^p \end{pmatrix} \quad . \tag{16}$$

We define protein expression noise as the coefficient of variation, i.e. as the ratio of the standard deviation divided by mean, of protein molecule numbers

$$\eta_p = \frac{\sigma_p}{[p]} \quad . \tag{17}$$

Calculating the noise in protein expression by solving the fluctuation dissipation theorem (equation (11)) for the variance in protein expression yields

$$\begin{aligned} \eta_p &= \sqrt{\frac{Var([p])}{[p]^2}} \\ &= \sqrt{\frac{1}{[p]} + \frac{1}{[m]\cdot(1-R\cdot S)} \cdot \frac{\lambda^p}{\lambda^p+\lambda^m} + \frac{1}{[\mu^T]} \cdot \frac{\lambda^p\cdot\lambda^m\cdot(\lambda^m+\lambda^p+\lambda^\mu)}{(\lambda^p+\lambda^m)\cdot(\lambda^m+\lambda^\mu)\cdot(\lambda^p+\lambda^\mu)} \cdot \left(\frac{R}{1-R\cdot S}\right)^2} \end{aligned} \tag{18}$$

18

Noise in protein expression can be understood as a mixture of noise arising from intrinsic and extrinsic sources (Elowitz et al., 2002, Swain et al., 2002). Here we consider the first two terms in equation (18) (describing noise arising from protein and mRNA molecule numbers) as intrinsic noise, while we consider the third term (describing noise arising from microRNA molecule numbers, the external regulator) as extrinsic noise.

Fluctuations in molecule numbers are a consequence of the inherent stochasticity of all molecular processes, here in particular the birth and death processes of mRNA and protein molecules. In our model description intrinsic noise consists of noise at the protein level, which is inversely related to the number of protein molecules, and noise in the number of mRNA molecules which propagates to the protein level (Paulsson, 2004, Pedraza and van Oudenaarden, 2005). How strongly noise propagates from the mRNA to the protein level depends on the time scales on which the protein and the mRNA fluctuations dissipate.

Re-writing intrinsic noise in terms of the average number of proteins that would be produced from an mRNA throughout its life-time when not microRNA regulated $b_0 = k^p/d^m$ yields

$$\eta_{int} = \sqrt{\frac{1}{[p]} \cdot \left(1 + \frac{b_0}{\frac{(1-R \cdot S)^2}{1-R} + \frac{d^p}{d^m} \cdot (1 - R \cdot S)} \right)} \quad . \tag{19}$$

For most genes the number of proteins produced per mRNA lifetime is much larger than one ($b_0 \gg 1$) and the constitutive degradation of the mRNA is much faster than that of the protein ($d^m \gg d^p$) (Schwanhäusser et al., 2011). Equation (19) then simplifies to

$$\eta_{int} = \sqrt{\frac{b_0}{[p]} \cdot \frac{1 - R}{(1 - R \cdot S)^2}} \quad . \tag{20}$$

Intrinsic noise of an unregulated gene ($R = 0$) declines with increasing protein expression and its size is determined by the proteins produced per mRNA lifetime (as discussed in chapter 1). MicroRNA regulation ($R > 0$) reduces intrinsic noise, but the degree of reduction is smaller if the microRNA pool is saturated by regulated mRNAs (Figure 8A). However, in the experiments presented in the next Chapter (ref) we find that saturation of microRNA regulation only happens at high expression levels where intrinsic noise is negligible. This is further corroborated by estimates of microRNA-mRNA dissociation constants, which govern saturation, as $K > 0.1 \, nM$, what corresponds to greater than 10^2 mRNA molecules in mammalian cells (Wee

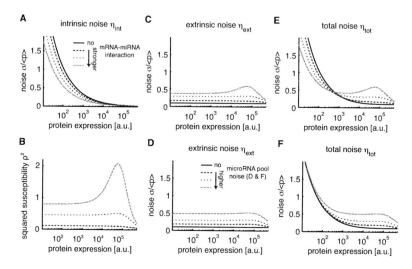

Figure 8: **Effects of microRNA regulation on noise in protein expression**
(A) Intrinsic noise of an unregulated gene (black) declines with increasing protein expression. MicroRNA regulation reduces intrinsic noise as a function of repression strength (gray shades and dashed lines, varied by the mRNA-microRNA interaction strength as in Figure 7).
(B) Susceptibility of mRNA levels towards changes in microRNA levels is constant across low expression, but can be amplified if the microRNA pool gets saturated in cases of strong mRNA-microRNA interaction.
(C) Additional extrinsic noise - propagating from microRNA pool noise - is constant across expression levels, but can be amplified when the microRNA pool gets saturated, especially for higher mRNA-microRNA interaction strength (here, microRNA pool noise is fixed).
(D) The effects of higher microRNA pool noise on additional extrinsic noise (here, mRNA-microRNA interaction strength is fixed to second strongest value used in (C)).
(E) Total noise is reduced by microRNA regulation at low expression, due to less intrinsic noise, but increased at high expression, due to propagation of microRNA pool noise. Both effects are more pronounced for higher mRNA-microRNA interaction strength.
(F) The effects of higher microRNA pool noise on total noise (mRNA-microRNA interaction strength fixed to second strongest value used in (E))

et al., 2012).

Assuming $S \approx 0$, the reduction of intrinsic noise in a microRNA-regulated gene compared to an unregulated gene (at equal protein expression) is therefore

$$\frac{\eta_{int,reg}}{\eta_{int,unreg}} \approx \sqrt{\frac{1}{1-R}} = \sqrt{\frac{[p_0]}{[p]}} = \sqrt{r} \quad , \tag{21}$$

the square root of the fold-repression induced by microRNA regulation r, which is consistent with predictions from similar models investigating the noise effects of small RNA in bacteria and microRNAs (Mehta et al., 2008, Noorbakhsh et al., 2013). Our model therefore predicts that the microRNA-mediated intrinsic noise reduction is independent of the molecular details of the mRNA-microRNA interaction, i.e. the affinity or sequence complementarity of the microRNA to the target site or how abundant the microRNA is expressed. These factors only affect intrinsic noise reduction as far as they alter the repression of protein expression.

Counteracting the reduction of intrinsic noise, the pool of microRNA molecules acts as an additional extrinsic noise source. This additional noise is described by the third term in equation (18), which we re-write as

$$\eta_{ext}^2 = \eta_\mu^2 \cdot \tau_{\mu,m,p} \cdot \varphi^2 \quad . \tag{22}$$

The propagation of noise in the pool of microRNA molecules is governed by the susceptibility of mRNA levels towards changes in the microRNA $\varphi = \frac{R}{1-R\cdot S}$. The susceptibility is understood as the relative change in mRNA levels upon a relative change in the microRNA levels. It is determined by the repression conferred by the microRNA but can be amplified if the microRNA pool becomes saturated by the regulated mRNAs (Figure 8B+C) (Paulsson, 2004, Mehta et al., 2008, Noorbakhsh et al., 2013).

Further, the propagation of noise on the microRNA level to the protein level is attenuated by the time-averaging of fluctuations between the microRNA, mRNA and protein levels (with the assumption of faster relaxation time on the mRNA level than on the microRNA and protein level ($\lambda^m \gg \lambda^\mu, \lambda^p$) (Schwanhäusser et al., 2011, Baccarini et al., 2011)):

$$\tau_{\mu,m,p} = \frac{\lambda^p \cdot \lambda^m \cdot (\lambda^m + \lambda^p + \lambda^\mu)}{(\lambda^p + \lambda^m) \cdot (\lambda^m + \lambda^\mu) \cdot (\lambda^p + \lambda^\mu)} \approx \frac{\lambda^p}{\lambda^p + \lambda^\mu} \in [0,1] \quad . \tag{23}$$

For further analysis, we define the *microRNA pool noise* $\tilde{\eta}_\mu$ as the noise that is 'seen' by the regulated gene as

$$\tilde{\eta}_\mu^2 = \eta_\mu^2 \cdot \tau_{\mu,m,p} \quad , \tag{24}$$

which is therefore a lower bound on the actual noise in the microRNA pool η_μ. Consequently, higher microRNA pool noise increases the additional extrinsic noise in a microRNA regulated gene (Figure 8D).

The total noise in protein expression (equation (18)) in its most approximated form can then be re-written as

$$\eta_{p,total} = \sqrt{\frac{b_0}{[p]} \cdot \frac{1}{r} + \tilde{\eta}_\mu^2 \cdot \varphi^2} \quad . \tag{25}$$

The combined effects of intrinsic noise reduction and additional extrinsic noise will lead to reduced total noise levels at low expression levels, where intrinsic noise dominates, but will result in increased noise at high expression levels, where intrinsic noise is diminished (Figure 8E). The size of both effects will increase with repression strength. Furthermore, increases in additional extrinsic noise due to higher microRNA pool noise will likely only affect expression levels where intrinsic noise is small, due to total noise being the square root of the summed squared of intrinsic and extrinsic noise ($\eta_{total} = \sqrt{\eta_{int}^2 + \eta_{ext}^2}$) (Figure 8F).

In summary, the mathematical model of microRNA-mediated noise effects makes precise, testable predictions. First, it predicts that intrinsic noise should be reduced by the square root of fold-repression. Second, it predicts that microRNA pool noise propagates to the regulated gene, also depending on the strength of repression. Third, therefore the effects of microRNA regulation on total noise should be reduced noise at low expression, but increased noise at high expression. However, it is not per se clear, what cellular expression levels of genes correspond to the here described low and high expression regimes, where microRNA regulation will have opposite effects on noise in protein expression.

2.3 Experimental validation of microRNA noise effects

In this chapter I present the experimental validation of the model predictions about microRNA noise effects derived in the previous chapter. I use a single cell fluorescent reporter setup to probe microRNA noise effects in mouse embryonic stem cells. First I test the general agreement of the data with microRNA effects on overall noise. A modified setup is then used to specifically test intrinsic noise effects. Finally I explore noise in the regulating microRNA pools and how it is affected by mixing of different microRNA species. The analysis present in this chapter will be published in Schmiedel et al. (2015). Yannan Zheng helped me to establish the reporter setup in mouse embryonic stem cells. Sandy Klemm contributed the statistical model used to deconvolute flow cytometry noise measurements.

2.3.1 Experimental validation of noise effects for miR-20a in mouse embryonic stem cells

We devised an experimental setup in order to test the predictions from our noise model as presented in the previous chapter.

To quantify protein levels and their fluctuations we adapted a dual fluorescent reporter system (Mukherji et al., 2011), where two different reporters (ZsGreen and mCherry) are transcribed from a common bi-directional promoter (Figure 9A). Zs-Green fluorescence intensity serves as the control of transcriptional activity from plasmids in each cell. mCherry fluorescence intensity is used to investigate noise in protein expression. It can be modified to contain varying numbers and types of microRNA binding sites in its 3'untranslated region (3'UTR) to probe microRNA-dependent noise effects.

For each experiment we performed three separate transfections of mouse embryonic stem cell cultures. One, a mock transfection, using a plasmid that does not code for fluorescent proteins, to determine cellular autofluorescence. Two, the transfection of the proper reporter plasmid with an empty mCherry 3'UTR to inform about unregulated protein expression noise (referred to as control reporter). Three, the transfection of the proper reporter plasmid with a particular miRNA binding site configuration in the mCherry 3'UTR to inform about microRNA-mediated noise effects. Single cell fluorescence was quantified using a flow cytometer after mESC cultures were grown for 48 hours (Figure 9B, for details on experiments see section 4.2). For the mESC cultures transfected with control and microRNA-modified plasmid variants we obtained on average $\sim 10^5$ measurements of cells above autofluorescence

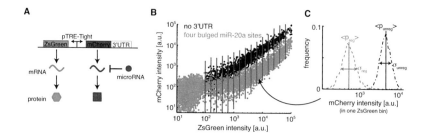

Figure 9: **Experimental setup to quantify microRNA-dependent noise effects in mouse embryonic stem cells**

(**A**) Plasmid reporter system with two genes coding for fluorescent proteins, ZsGreen and mCherry, transcribed from a common bi-directional promoter. mCherry 3'UTR can be modified to contain microRNA binding sites.

(**B**) Overlay of two flow cytometry measurements of mESC populations transiently transfected with different variants of the plasmid system: empty mCherry 3'UTR (black) and mCherry 3'UTR containing four bulged miR-20a binding sites (blue). For further processing cells are binned according to ZsGreen intensity (red lines) and cells below ZsGreen background are discarded (grey)

(**C**) Mean and noise (standard deviation divided by mean) of mCherry intensities are calculated from marginal distributions in each bin.

background. ZsGreen fluorescence intensity, which varied roughly three orders of magnitude between cells (mostly due to differential uptake of plasmid copies per cell), was then used to bin cells with similar transcriptional activity.

We are interested in the mean and noise of cellular mCherry fluorescence intensities in each bin, in order to investigate microRNA-dependent noise effects. However, the measured fluorescence intensities are mixtures of the actual mCherry fluorescence intensities and cellular autofluorescence intensities. Accurate estimates of bin-wise mean mCherry fluorescence intensities can be obtained by subtracting mean autofluorescence, which we calculate from mock transfections. Obtaining accurate estimates of the noise of mCherry fluorescence intensities is more complicated. Processes correlating mCherry fluorescence and cellular autofluorescence, such as varying cell sizes, inflate the measured noise. Furthermore, noise introduced by the measurement itself have to be accounted for. My colleague Sandy Klemm devised a comprehensive statistical framework that we used to deconvolute noise in mCherry intensities from flow cytometry measurements (see section 4.3).

To learn about microRNA-mediated noise effects, we constructed three plasmid

24

variants with different binding site configurations in the mCherry 3'UTR for miR-20a, a microRNA endogenously expressed in mESC. The three variants are: one bulged miR-20a binding site, one perfect miR-20a binding site, and four bulged miR-20a binding sites (for details see section 4.2). The perfect binding site has full complementary across all 21 nucleotides of miR-20a, which elicits cleavage of the mRNA through catalytic activity of Argonaute proteins (Haley and Zamore, 2004). Bulged binding sites have full complementarity except for mismatches at positions 10-12 in the miR-20a sequence (creating the bulge). This precludes cleavage and repression is instead elicited via interaction of protein co-factors with the polyA tail as well as the translation process (Hutvágner and Zamore, 2002, Haley and Zamore, 2004). These two types of high complementary binding sites were chosen to achieve strong repression and distinct noise effects. The next chapter shows that results presented here are not limited in their generality by these differences to endogenous-style binding sites.

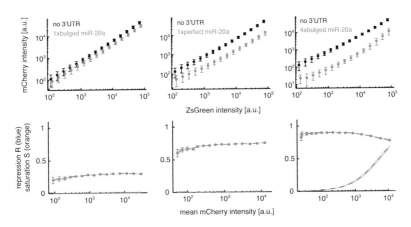

Figure 10: **Effects of miR-20a regulation on mean mCherry fluorescence intensities**
(Upper row) Bin-wise mean (dots) and standard deviation (error bars) of mCherry intensities for the three different miR-20a regulated reporter variants compared to an unregulated control reporter without miR-20a binding sites.
(Lower row) Repression R (blue) and saturation S (red), obtained from model fits to mean mCherry intensities, as a function of mean mCherry intensities.

First we compared the mean mCherry intensities of miR-20a regulated reporters to the unregulated control reporter. As expected, miR-20a regulation leads to repression of mCherry expression and repression is stronger when mCherry is regulated via a perfect or several bulged binding sites (Figure 10).

We estimated the mean parameters repression R and saturation S by fitting our model to the bin-wise mean mCherry intensities (procedure of model fits is described in section 4.2). The fits show that repression stays constant across the whole range of mCherry intensities for the 1xbulged and 1xperfect miR-20a reporter constructs. For the 4xbulged miR-20a reporter construct repression is constant across low and intermediate expression ranges but falls off at high expression and the model fits indicated that this is due to increasing saturation of miR-20a (Figure 10). These results are in line with a previous study that investigated the repression and saturation effects of miR-20a in HeLa cells (Mukherji et al., 2011).

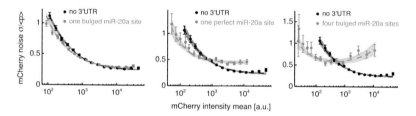

Figure 11: Effects of miR-20a regulation on noise in mCherry protein expression
Bin-wise noise of mCherry intensity as a function of mean mCherry intensity for the three different miR-20a regulated reporters (blue) compared to respective unregulated control reporters (black). Panels are ordered from left to right according to increasing repression of mCherry by miR-20a. Dots: data, lines and shaded area: model fit.

Next we investigated noise in mCherry intensities as a function of mean mCherry intensities (Figure 11). The data show that noise in miR-20a regulated mCherry reporters is smaller at low mean expression and larger at high mean expression compared to unregulated mCherry control reporters at equivalent mean expression levels.

We fitted the noise model to the data. The noise model fit to the unregulated control reporter contains the term for intrinsic noise as well as a constant term for microRNA-independent extrinsic noise. The noise model fit to the regulated reporters contain the term for intrinsic noise modified by the estimated mean parameters repression and saturation, the constant term for microRNA-independent extrinsic noise as well as the microRNA-dependent extrinsic noise. The noise model fits to the regulated reporters therefore only have one free microRNA-dependent parameter, the microRNA pool noise $\bar{\eta}_\mu$. The model fits yield accurate agreement with the experimentally observed total noise profiles and thus give a first indication that our mathematical model is also able to quantitatively describe microRNA-mediated noise effects.

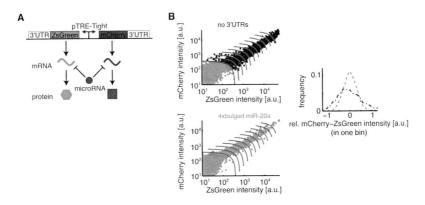

Figure 12: **Experimental setup to investigate microRNA-dependent intrinsic noise effects**
(**A**) Modified plasmid reporter system where the 3'UTRs of both fluorescence reporter genes are modified to contain an identical set of miR-20a binding sites.
(**B**) Example of flow cytometry measurements of reporters with either no binding sites (upper panel, black) or four bulged miR-20a binding sites (lower panel, blue) in the 3'UTRs of both ZsGreen and mCherry. Binning was performed along summed ZsGreen and mCherry intensities (indicated by red lines). Right panel shows an example of bin-wise relative intensity differences of both reporter that were used to calculate intrinsic noise.

We set out to test in more detail whether the quantitative model predictions about noise effect sizes for both intrinsic and extrinsic noise are correct. To distinguish between microRNA-mediated intrinsic and extrinsic noise effects experimentally, we modified the plasmid reporter system such that both ZsGreen and mCherry contain identical 3'UTRs (Figure 12A). Now intracellular differences in the expression of both reporters can only result from processes individual to the expression of each gene, i.e. the processes that constitute intrinsic noise. To estimate intracellular differences in the expression of ZsGreen and mCherry, we binned cells along the summed ZsGreen and mCherry intensities. Bin-wise intrinsic noise was then calculate as the

27

normalized standard deviation of relative differences in ZsGreen and mCherry intensities (Figure 12B).

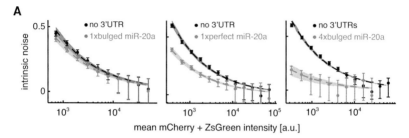

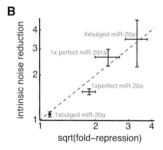

Figure 13: **MicroRNA-mediated intrinsic noise reduction**
(**A**) Intrinsic noise as a function of summed mean intensities for three different miR-20a bi-regulated constructs. Dots: data, lines and shaded area: model fit.
(**B**) Measured intrinsic noise reduction for bi-regulated reporter constructs compared to fold-repression, as measured independently for mCherry-regulated reporters (Figure 10). Error bars indicate standard deviation of estimates obtained from at least three biological replicates.

The profiles of intrinsic noise as a function of summed mean intensities show that microRNA regulation indeed reduced intrinsic noise as a function of repression strength (Figure 13A). To inform about the agreement of intrinsic noise profiles with the noise model the intrinsic noise data of each bi-regulated construct together with the unregulated construct were fit to a simplified model equation of the form $\eta_{intr} = \sqrt{\frac{X}{[Z+M]}}$. Here $[Z + M]$ is the summed mean intensities of ZsGreen and mCherry, as taken from the data, and X is a free scaling parameter to convert fluorescence intensities to molecule numbers.

The fits validate the model predictions for intrinsic noise. First, the model predicts intrinsic noise to decrease with expression in a $\sqrt{\frac{1}{n}}$-relationship, with n as the number of molecules present, which is validated by the good agreement between data and model fits for regulated as well as unregulated constructs (Figure 13A). Second, the model predicts that the intrinsic noise reduction should be as large as the square root of fold-repression conferred by the microRNA. We calculate the effect size of microRNA-mediated intrinsic noise reduction as the ratio between the fitted scaling parameters X for bi-regulated and unregulated reporter constructs. We find a good agreement between these estimates for intrinsic noise reduction and the square root of fold-repression, as observed for the miR-20a regulated mCherry reporters, thus validating the model prediction about the effect size of microRNA-mediated intrinsic noise reduction (Figure 13B).

We deduce that the additional noise observed in the total noise data must originate from extrinsic noise. We extracted extrinsic noise profiles from the total noise data (see Figure 11) by subtracting the model-inferred intrinsic noise portions (Figure 14A). Extrinsic noise of the unregulated mCherry reporter is constant across the whole expression range, indicating that noise propagating from external sources is independent of the mean expression of the reporter and thus justifying the additive constant extrinsic noise term in our model equations.

Extrinsic noise of each of the miR-20a regulated mCherry reporters is constant over the expression range and larger for the constructs with stronger miR-20a mediated repression. Extrinsic noise of the four bulged miR-20a binding site reporter shows an increase at high expression levels, which is consistent with extrinsic noise being amplified due to saturation of miR-20a regulation by the reporter mRNAs (see Figure 8 and Figure 10).

The noise model further predicts that additional microRNA-mediated extrinsic noise is a function of the underlying microRNA pool noise and its propagation to the regulated gene, which is dependent on repression and saturation. We would therefore expect that, when accounting for the strength of propagation, the model-based estimates of miR-20a pool noise from the three different reporter constructs should be identical. Fits of the full noise model to the total noise profiles, with the microRNA pool noise $\bar{\eta}_\mu$ as the only free microRNA-dependent parameter, indeed give within error equal estimates of miR-20a pool noise for all three mCherry reporters (Figure 14B).

In summary, the experiments presented here validate the model predictions that microRNA regulation reduces intrinsic noise while acting as an additional extrinsic noise source. The experiments also validate the predictions about the effect sizes of

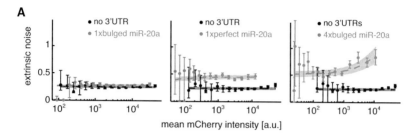

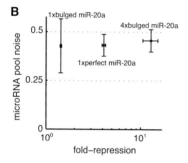

Figure 14: **MicroRNAs act as an additional extrinsic noise source**

(**A**) Extrinsic noise, extracted from total noise profiles, as a function of mean mCherry expression. Dots: model-based extracts of extrinsic noise, lines and shaded area: model fit of extrinsic noise terms to data.

(**B**) microRNA pool noise estimates from total noise profiles of the different miR-20a regulated mCherry reporters. Error bars indicate standard deviation of estimates obtained from at least three biological replicates.

intrinsic noise reduction and the propagation of microRNA pool noise to the regulated gene. Furthermore, the fact that the effect size do not depend on the exact composition of the miR-20a binding site, which elicit their repression through different regulatory modes, shows that microRNA-mediated noise effects do not depend directly on the specific molecular details of the mRNA-microRNA interaction. Instead these molecular details only affect effect sizes through altering the macroscopic mean parameters repression and saturation.

2.3.2 Generality of intrinsic noise reduction for microRNAs and other post-transcriptional repressors

Since the noise effect size relationships do not depend on the exact composition of miR-20a binding sites we would expect them to hold also for other microRNAs that use binding sites with completely different sequences. Furthermore, we would expect that different microRNA expression levels alter noise effect size only to the extend that they affect mean repression.

To test these hypotheses experimentally, we constructed eight additional reporters with mCherry 3'UTRs containing a perfect binding site for a variety of microRNAs that are endogenously expressed in mESC (Figure 15A). microRNAs were selected from a range of expression that yielded discernible noise effects compared to unregulated control reporters and perfect target sites were chosen to ensure optimal model fit estimates of effect sizes from the noise data.

The total noise profiles of these additional mCherry reporters exhibit the expected noise effect characteristics, namely reduced noise at low expression and increased noise at high expression, and show good agreement with model fits (Figure 15B), thus indicating that effect size relationships are met. We further validated this in two ways. First, we constructed a bi-regulated reporter construct to directly measure intrinsic noise reduction (as described in Figure 12) for one of the additional microR-NAs, miR-291a. The bi-regulated miR-291a reporter construct exhibited intrinsic noise reduction compared to the unregulated control reporter as large as the square root of fold-repression conferred by miR-291a (Figure 13B), in line with results for the miR-20a regulated reporters . Second, we used fits of a modified noise model, using the effect size of intrinsic noise reduction as a free fit parameter, to the total noise profiles. These fits independently confirm that intrinsic noise reduction equals within error the square root of fold-repression for all assayed microRNA reporters, even though repression by microRNAs varies across an order of magnitude (Figure 16A).

If the noise effects elicited by microRNAs do not depend on the exact details of molecular interaction or the levels of microRNAs, do they also generalize to other post-transcriptional repressors? microRNAs and their associated protein co-factors are special post-transcriptional regulators in the sense that they facilitated an interaction through a direct RNA-RNA base pairing, as opposed to most other post-transcriptional regulatory proteins, which interact with mRNAs through protein domains with affinities for specific RNA sequence motifs. However these differing details of the molecular interaction again should not change the noise effects resulting

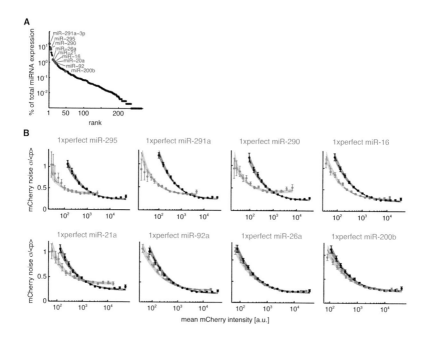

Figure 15: **A wide array of mESC microRNAs shows expected noise effects**
(**A**) Expression levels of mESC microRNAs as measured in Marson et al. (2008).
The microRNAs investigated are noted in red.
(**B**) Total noise profile for reporters with a perfect microRNA target sites. Black -
unregulated control reporters. Blue - reporters with perfect target sites for indicated
microRNAs. Dots: data, lines and shaded area: model fit.

from post-transcriptional repression in general. In order to test the generality of noise
effects for other post-transcriptional repressors we constructed a reporter with seven
AU-rich elements in the mCherry 3'UTR. AU-rich elements are sequence elements
used by a variety of RNA-binding proteins, which are thought to predominantly elicit
mRNA destabilization (Barreau et al., 2005). In agreement with our hypothesis the
total noise profile of the AU-rich reporter show the expected noise effects (Figure
16B), and intrinsic noise reduction is estimated to be close to the square root of
fold-repression (Figure 16A).

Finally, we investigated how the two different modes of post-transcriptional repression affect intrinsic noise reduction in more detail. MicroRNAs and other post-

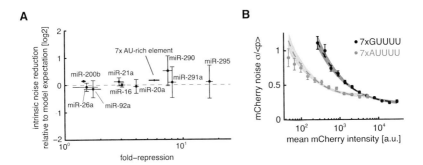

Figure 16: **Generality of intrinsic noise reduction for microRNAs and other post-transcriptional repressors**
(**A**) Comparison of model-based estimates of intrinsic noise reduction from total noise profiles to model predicted effect size for reporters with perfect microRNA target sites and AU-rich elements. Error bars indicate standard deviation of estimates from at least three biological replicates.
(**B**) Total noise profile for mCherry with seven AU-rich elements (blue). Unregulated control is a reporter with seven GUUUU-elements in the mCherry 3'UTR (black).

transcriptional regulators can repress protein production via mRNA destabilization or translational inhibition. The question arises whether intrinsic noise is affected differently by these two modes of repression. An extended model, allowing for varying contributions of mRNA destabilization and translational inhibition to overall post-transcriptional repression (see section **??**), predicts that both modes of repression should contribute equally to intrinsic noise reduction as far as they contribute to repression. While we did not directly measure contributions of both repressive modes in our reporter system, circumstantial evidence supports this notion. First, the mechanisms of post-transcriptional repression are known to differ dependent on the complementary of microRNA binding sites. Perfect complementary of microRNAs to mRNAs elicits cleavage of the mRNA through catalytic activity of Argonaute proteins, but imperfect complementarity precludes cleavage and repression is instead elicited via interaction of protein co-factors with the polyA tail as well as the translation process (Hutvágner and Zamore (2002), Haley and Zamore (2004) and data presented in chapter 2.5). While the first therefore mainly acts through mRNA destabilization, the latter will have additional contributions from translational inhibition. Nonetheless we observe no differences in the effect size of intrinsic noise reduction for the assayed reporters containing either perfect or bulged microRNA binding sites (Figure 13B and 16A). This is further supported by data for reporters with long 3'UTR stretches containing endogenous, low-complementary microRNA binding sites (Figure 18). Second, RNA-binding proteins acting through AU-rich elements are though to predominantly destabilize mRNAs and we find intrinsic noise reduction effects in agreement with our model. Furthermore, in chapter 2.5 I present data showing that in mouse embryonic fibroblasts regulation via AU-rich elements likely results in a combination of mRNA destabilization and translational promotion. While we have no data on the exact regulatory contributions for AU-rich elements in mESC, intrinsic noise reduction due to a net repression that results from opposed action of both regulatory modes is in no contradiction to our noise model and our experimental results.

In summary, the analyses presented here support the plausibility that reduction of intrinsic noise is a generic property of all post-transcriptional repressors.

2.3.3 Additional extrinsic noise from individual and mixed microRNA pools

Additional extrinsic noise results from the propagation of noise in the microRNA pool. While propagation, like intrinsic noise reduction, is governed by the mean parameters repression and saturation and should thus not crucially depend on the molecular details of the mRNA-microRNA interaction, the magnitude of noise might differ between individual microRNA pools.

Results presented in Figure 14B validate that our noise model is able to reliable estimate microRNA pool noise. We thus used model fits to total noise profiles to extract microRNA pool noise for all nine perfect target sites reporters. We find that pool noise differs between different microRNAs and on average decreases as a function of repression conferred by the microRNAs on mCherry reporters (Figure 17A).

Noise in microRNA pools should be thought of as fluctuations in the availability of functional microRNA-Argonaute complexes. It might therefore originate from two sources: intrinsic noise in the number of molecules of specific microRNAs and noise in the overall levels of Argonaute proteins (and other co-factors). These two sources should crucially differ in two aspects. First, intrinsic noise in the number of molecules of a specific microRNA should decrease with increasing availability of functional microRNA-Argonaute complexes. On the contrary noise from the overall levels of Argonaute proteins should be constant with increasing availability of functional microRNA-Argonaute complexes, because their overall mean levels are fixed. Second, intrinsic noise in the number of individual microRNAs should be uncorrelated between different microRNA pools, while noise from the overall levels of

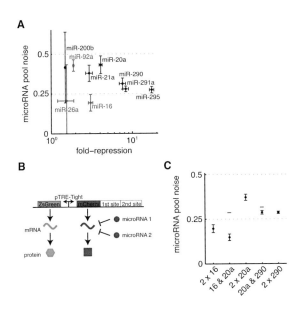

Figure 17: **MicroRNA pool noise of mESC microRNAs and its attenuation in mixed microRNA pools**

(**A**) Model-based estimate of microRNA pool noise for assayed mESC microRNAs as a function of mCherry repression. MicroRNAs with one genomic copy are indicated in black, those with two genomic copies are indicated in red. Error bars indicate standard deviation of estimates from at least three biological replicates.

(**B**) Experimental setup to compare microRNA pool noise between individual and mixed microRNA pools. mCherry is modified to contain two perfect binding sites, either for the same or for two different microRNAs.

(**C**) Comparison between noise in individual ('2x...') and mixed ('...&...') microRNA pools. Red dashes in mixed pool columns indicate expectation for pool noise if noise of the respective individual pools were fully correlated.

Argonaute proteins should be correlated between different microRNA pools.

The observation that microRNA pool noise decreases with conferred repression but levels off for the most highly expression mESC microRNAs (see Figure 15A) thus indicates that microRNA pool noise is a mixture of intrinsic noise of individual microRNAs and noise from Argonaute proteins and other cofactors. However, two microRNAs, miR-16 and miR-26a, do not obey the general trend and instead show low pool noise levels even at low conferred repression. We subsequently found that these two microRNAs, as well as miR-92a, are transcribed from two instead of one independent genomic loci (Figure 17A, indicated in red). This is therefore an indication that intrinsic noise in the production of individual microRNAs is a prominent contributor to microRNA pool noise, since intrinsic noise in the production from two independent loci should be uncorrelated and therefore average out.

To test this hypothesis, we constructed reporters with two perfect binding sites in the mCherry 3'UTR (Figure 17B). We compared reporters with perfect target sites for miR-20a and either miR-16 or miR-290 to reporters with two perfect target sites for miR-16, miR-20a or miR-290, respectively. We indeed find that noise levels in the mixed pools are lower than expected if the individual microRNA pools were fully correlated and can even be lower than the noise in the individual microRNA pools (Figures. 17C).

Our data therefore suggest that microRNA pool noise is made up of intrinsic noise in individual microRNA levels as well as general noise from Argonaute proteins and other co-factors. However, the contributions from intrinsic noise in the number of microRNA molecules can be mitigated when pools consist of independently transcribed microRNAs, therefore leading to decreased overall pool noise.

2.4 microRNA noise effects in endogenous target situations

In this chapter I address the question whether the findings from our reporter system are applicable to endogenous target situations. I validate noise effects for full 3'UTRs of several known microRNA targets, test how the noise effects relate to endogenous expression levels and finally investigate how many genes are substantially regulated by microRNAs in mESC. The analysis present in this chapter will be published in Schmiedel et al. (2015). Yannan Zheng contributed the Casp2, Lats2, and Rbl2 3'UTR reporter plasmids. Stefan van der Elst, Mauro Muraro, Dylan Moojiman, and Lennart Kester helped with FACS experiments and downstream RNA-sequencing.

Three issues arise when we try to extrapolate the fundamental microRNA-mediated noise effects to understand the potential of microRNA regulation to affect protein expression noise of genes *in vivo*. First, how are microRNA noise effects re-shaped by the architecture of endogenous mRNA-microRNA interactions, where mRNAs usually only contain low complementary binding sites, albeit for many different microRNAs at once? Second, how do microRNA noise effects map to the expression ranges of their endogenous targets? And lastly, how many genes might actually experience substantial microRNA noise effects *in vivo*?

Contrary to the reporter setups with high complementary microRNA binding sites used in the previous chapter, mRNAs usually only contain low complementary microRNA binding sites. Endogenous interactions are mainly governed by the seed region (microRNA 5' positions 2-7) with occasional additional complementarity in 3' flanking regions (Friedman et al., 2009). Repression resulting from individual low complementary mRNA-microRNA interactions is usually modest (below twofold) (Lewis et al., 2005, Baek et al., 2008, Selbach et al., 2008). However, mRNAs often have 3'UTRs longer than a thousand nucleotides that contain numerous binding sites for various microRNAs, several of them usually co-expressed in diverse cellular states (Enright et al., 2003, Krek et al., 2005).

Building on the results presented in the previous chapter we thus want to understand whether many weak interactions can lead to substantial microRNA-mediated noise effects and how the combinatorial nature of targeting affects additional extrinsic noise of regulated genes. In order to capture endogenous mRNA-microRNA interactions in an unbiased way, we decided to assay them in their natural context, i.e. in their native 3'UTRs. We selected the 3'UTRs of four genes that are well known microRNA targets in mouse embryonic stem cells. The 3'UTR of Casp2 has a length of 2,019 nucleotides and is predicted to contain 51 binding sites for mESC microR-

NAs. The 3'UTR of Lats2 has a length of 1,605 nucleotides and 49 predicted mESC microRNA binding sites. The 3'UTR of Rbl2 has a length of 1,389 nucleotides and 32 predicted mESC microRNA binding sites. Additionally, we assayed a 3'UTR stretch of Wee1 comprising nucleotides 130 to 610, which contains 25 predicted mESC microRNA binding sites.

These 3'UTRs were cloned behind mCherry. In order to isolate microRNA-mediated effects from other regulatory effects of the 3'UTRs, we constructed control reporters where the 3'UTRs have point-mutations in computationally predicted microRNA binding sites to abolish their repressive effects (see section 4.2).

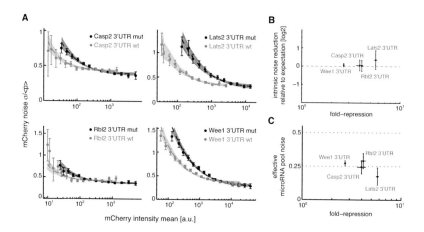

Figure 18: **Noise effects for 3'UTRs of endogenous microRNA targets**
(**A**) Total noise profiles for reporters with 3'UTRs of endogenous microRNA targets cloned behind mCherry. Effects of wild-type 3'UTRs (blue) are compared to control 3'UTRs with point mutations in computationally predicted mESC microRNA binding sites.
(**B**) Model-fit based estimates of intrinsic noise reduction from total noise profiles compared to expectation from observed repression. Error bars indicate standard deviation of intrinsic noise reduction estimates and observed repression from three biological replicates.
(**C**) Estimates of microRNA pool noise from total noise profiles for the mixed microRNA pools regulating the wild-type 3'UTRs. Error bars indicate standard deviation of pool noise estimates and observed repression from three biological replicates.

mCherry reporters with the wild-type 3'UTRs exhibit 3 to 5.5 fold repression compared to the control reporters with mutated 3'UTRs (Figure 18B&C). As expected from such strong repression, total noise profiles show substantial reduction in noise at low expression (Figure 18A) and model fits estimate the intrinsic noise reduction to be within error equivalent to the square root of fold-repression (Figure 18B). On the contrary, increases in total noise are either not observed over the range assayed by our reporter (Lats2 3'UTR) or surprisingly small given the strong repression of the 3'UTR reporters by microRNAs (Figure 18A). The model fits indicate that this is due to low noise in the mixed microRNA pools regulating the wild-type 3'UTRs (Figure 18C). This is thus consistent with our previous finding that mixing of individual microRNA pools can average out independent fluctuations (Figure 17). Additionally, it highlights another effect of microRNA pool noise: given a certain microRNA-mediated repression, the size of the microRNA pool noise can effectively shift the balance between intrinsic and extrinsic noise effects. It therefore determines the gene expression level at which the net effect of microRNA regulation switches from reduced to increased total noise.

In summary, endogenous mRNA-microRNA interactions, comprised of mixtures of many weak binding sites for different microRNAs, obey the general noise effect size relationships as discussed in the previous chapter. They can lead to strong repression, therefore indicating that microRNA-mediated noise effects could play a substantial role *in vivo*. Furthermore, mixing of many microRNA-target interactions on 3'UTRs results in low microRNA pool noise. This in turn leads to a shift in the balance between intrinsic and extrinsic noise effects, extending the range of intrinsic noise reduction to higher expression levels as compared to individual microRNA-target interactions.

Considerations about shifting the balance between intrinsic and extrinsic noise effects for endogenous mRNA-microRNA interactions bring us back to the question already posted at the end of the model section: How do the microRNA-mediated noise effects relate to the expression of endogenous microRNA targets?

We set out to establish the relation between the expression range assayed by our fluorescence reporter system and the expression distribution of the mESC transcriptome. We used fluorescence activated cell sorting to collect four populations of cells expressing the reporter system at different ZsGreen intensities (Figure 19A) and subsequently isolated and sequenced RNA from each cell population. Comparison of population-wise mCherry fluorescent intensities to mCherry mRNA levels allowed us to calculate a fluorescence intensity to FPKM conversion factor (Figure 19B). When mapping the range covered by our reporter system to the transcriptome expression distribution of mESC, we find that our reporter assay covers the range

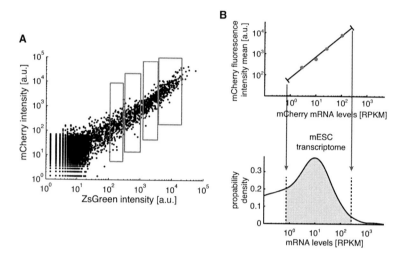

Figure 19: **Mapping the fluorescence reporter range to endogenous transcriptome expression**
(**A**) Fluorescence activated cell sorting was used to collect cell populations expressing the reporter at different levels (selection gates marked as gray boxes).
(**B**) RNA-sequencing of cell populations allows to calculate a conversion factor between mCherry fluorescence intensities and mCherry mRNA levels (upper panel). Lower panel shows how the range covered by the reporter system relates to the mESC transcriptome.

of 25% to 99% of expressed genes in mESC. This therefore indicates that the noise effects observed in our reporter assay are of relevance for endogenous microRNA targets.

We used the fluorescence intensity to FPKM conversion factor to extrapolate the noise effects from reporters with endogenous 3'UTRs to the mESC transcriptome (Figure 20A). For all four endogenous 3'UTR reporters, microRNA regulation results in reduced noise across most of the mESC transcriptome distribution. Noise reduction is maximal in the lower half of the mESC transcriptome distribution and extends beyond the 90th expression quantile, for the Lats2 3'UTR even close towards the 99th expression quantile. Furthermore, the mRNA expression levels of the assayed genes lie within the range of reduced total noise, indicating that they should experience microRNA-mediated noise reduction *in vivo*. However, note that a criti-

40

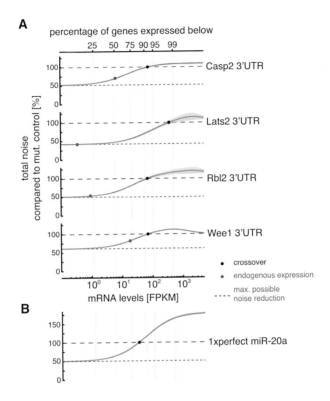

Figure 20: **Extrapolation of noise effects to mESC transcriptome expression levels**
(**A**) Model-based extrapolation of microRNA-mediated total noise effects for endogenous 3'UTR mCherry reporters. Indicated are FPKM values (lower x-axis) and percentage of genes expressed below a certain value (upper x-axis).
(**B**) microRNA-mediated total noise effects for mCherry reporter with perfect miR-20a binding site as reference to show how low noise in mixed pools extends the range of total noise reduction for endogenous 3'UTRs. Axis as in (**A**).

cal assumption for this deduction is that noise in the expression of the endogenous genes is similar to noise in our reporter system. In principle differing transcriptional dynamics, microRNA-independent mRNA half-lives, protein half-lives, and additional external regulators could lead to different balances of intrinsic and extrinsic noise in the protein expression of these genes, which could therefore also affect the balance of microRNA-mediated intrinsic and extrinsic noise effects.

A comparison of noise effects for the endogenous 3'UTR reporters to those of the reporter with a perfect miR-20a binding sites (with repression similar to Casp2 and Rbl2 3'UTRs) illustrates the influence of combinatorial targeting for the potential of mESC microRNAs to reduce noise of their targets (Figure 20B). Low noise in mixed microRNA pools results in an extended range of noise reduction and therefore includes hundreds of additional genes to experience potential microRNA-mediated noise reduction.

The question then arises how many genes in mESC are actually substantially regulated by microRNAs and at what levels those genes are expressed. As mentioned above, most individual mRNA-microRNA target interactions only lead to modest repression (Lewis et al., 2005, Baek et al., 2008, Selbach et al., 2008), which would therefore only result in small microRNA-mediated noise effects. However, the data from the endogenous 3'UTR reporters show that combinatorial regulation by mESC microRNAs can results in up to 5.5-fold repression. Are these genes lucky exceptions?

We used published data from a study comparing transcriptome expression between wild-type and microRNA-deficient Dicer knockout mESC to estimate the extend of combinatorial repression by mESC microRNAs (Leung et al., 2011). Calculating repression as the ratio of expression between knockout and wild-type cells shows that hundreds of genes are repressed more than two-fold and dozens of genes are repressed even more than four-fold (Figure 21A). Furthermore, most strongly repressed genes are found to be expressed at low to intermediate expression levels in wild-type mESC. This observation is consistent with results from previous studies that microRNAs preferentially target lowly expressed genes while selectively avoid highly expressed genes (Farh et al., 2005, Sood et al., 2006).

To increase the likelihood that expression differences between wild-type and microRNA-deficient mESC are indeed microRNA-mediated, we defined a set of high confidence microRNA targets. We calculated the intersection of genes that were found to be bound to Argonaute proteins (Leung et al., 2011) and genes predicted to have at least one conserved binding site for any of the microRNAs expressed in mESC, a set we termed 'supported' genes (predictions from Targetscan (Garcia et al., 2011), mESC microRNA expression from Marson et al. (2008)). Consistent with the notion that a major fraction of strong expression differences is due to microRNA regulation, we find that the 'supported' gene set is enriched for strongly repressed genes (Figure

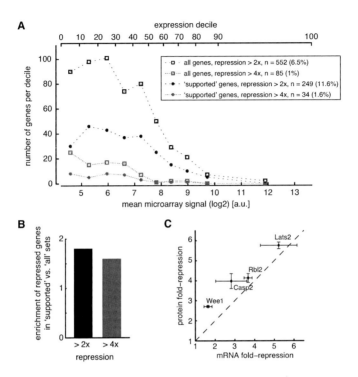

Figure 21: **Repression of genes by mESC microRNAs**

(**A**) Repression of genes in mESC as determined from expression differences in microRNA-deficient Dicer knockout to wild-type cells (Leung et al., 2011). Genes are binned according to expression deciles. 'Supported' set are genes found to be bound to Argonaute proteins and contain at least one conserved binding site for mESC microRNAs.

(**B**) Enrichment of strongly repressed genes in 'supported' gene set compared to set containing all expressed genes.

(**C**) Comparison of mRNA repression to protein repression as observed for endogenous 3'UTR reporters. Error bars indicate standard deviation of estimates of three replicated microarray measurements for mRNA repression and three biological replicates for reporter-based protein repression.

21B) and that the trend for those genes to be lowly expressed persists (Figure 21A). Furthermore, the calculated mRNA repression shows excellent agreement with the protein repression observed for our endogenous 3'UTR reporters (Figure 21C).

These data therefore show that combinatorial microRNA regulation results in substantial repression for hundreds of genes in mESC. And the finding that most strongly repressed genes are lowly expressed suggests they should therefore have reduced protein expression noise *in vivo*.

In summary, we conclude that the noise effects of microRNA regulation apply to endogenous microRNA targets. The common combinatorial regulation of endogenous targets by multiple microRNAs leads to strong intrinsic noise reduction and low microRNA pool noise, therefore extending the range of net noise reduction. We estimate that noise reduction for endogenous targets reaches beyond the 90th percentile of the mESC transcriptome distribution. Furthermore, hundreds of genes lowly expressed genes in mESC exhibit strong microRNA-mediated repression, thus suggesting they should have reduced noise due to microRNA regulation.

2.5   microRNAs predominantly increase production ratios

In the previous chapters I have shown the potential of microRNA regulation to reduce protein expression noise. The characteristic architecture of endogenous combinatorial microRNA regulation, the fact that therefore most expressed genes should experience a net reduction of protein expression noise and the observation that most strongly repressed genes are lowly expressed, make for a plausible argument that noise reduction is one of the biological functions microRNA regulation. In the following chapters I present further evidence consistent with this notion.

In Chapter 1 we have investigated the balance between transcriptional and post-transcriptional processes in the expression of genes in NIH3T3 cells, which led us to conclude that protein expression noise is controlled by production ratios (contributions from transcription rate versus post-transcriptional efficiencies). Noise-prone genes exhibit larger production ratios, which are in part mediated by lower post-transcriptional production efficiencies. It is therefore tempting to speculate that, if microRNAs had a prominent role in reducing protein expression noise, we should find signatures of increased microRNA regulation of noise-prone genes.

To test this hypothesis, we used Targetscan (Friedman et al., 2009) to predict conserved binding sites for microRNAs expressed in mouse embryonic fibroblasts (Landgraf et al., 2007) in the 3'UTRs of genes assayed by Schwanhäusser et al. (2011). These predictions show that genes with lower translation rates are enriched for microRNA binding sites and that, within groups of genes with similar translation rates, genes with short-lived proteins are enriched for microRNA binding sites over those with long-lived proteins (Figure 22A). Furthermore, transcription factors, which we also found to have larger production ratios than expected, show a strong enrichment for microRNA binding sites compared to all other genes (Figure 22B) (Enright et al., 2003).

However, this analysis falls short in establishing a direct link between microRNA regulation and increased production ratios, and enriched microRNA targeting for both sets of genes could be due to other reasons.

Two main ideas for the function of microRNAs have emerged over the years. The suppression of unwanted protein expression and the 'fine-tuning' of protein expression (Bartel and Chen, 2004), which we have shown to lead to reduced noise in protein expression. Both ideas are necessarily based on the fact that microRNAs reduce protein production post-transcriptionally. The crucial difference is in how microRNA regulation needs to be coordinated with the transcriptional level in order to

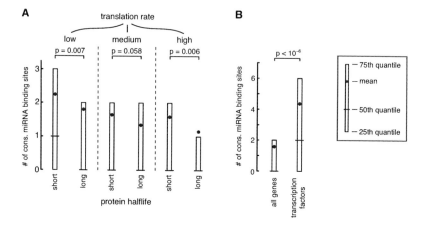

Figure 22: **Enriched microRNA regulation of genes with short-lived proteins and transcription factors**
(**A**) Number of conserved binding sites for microRNAs expressed in MEF cells for genes binned according to (1) their translation rate and (2) their protein half-life (same as in Figure 3B). p-values: two-sided Wilcoxon rank sum test for equal medians. Dot indicates mean, bar indicates median, boxes extend from lower to upper quartile (see box on right).
(**B**) Number of conserved binding sites for microRNAs expressed in MEF cells for transcription factors (GOid 3700) compared to all genes analyzed. p-value: two-sided Wilcoxon rank sum test for equal medians.

achieve each function (Figure 23). In order for microRNAs to suppress unwanted expression of target genes, we would expect these genes to also be repressed on the transcriptional level in the first place. On the contrary, for microRNAs to decrease noise, their action has to be linked to transcriptional activation in order to increase production ratios while keeping the mean protein expression constant. The key to understand whether microRNA regulation acts to decrease protein expression noise therefore lies in its coordination with transcriptional regulation.

We set out to investigate coordinated regulation of microRNA targets in the mouse embryonic fibroblast dataset (Schwanhäusser et al., 2011) by disentangling the contributions of transcriptional and post-transcriptional processes to their protein production (Figure 24A). We compared the set of genes predicted to be targeted by expressed microRNAs to control sets of genes which we matched for 3'UTR length, which is itself a major determinant of gene expression rates. First we established that

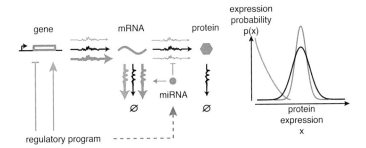

Figure 23: **The effects of coordinated regulation of transcription and microRNA regulation**
Depending on the sign of transcriptional regulation, coordinated regulation on the transcriptional and post-transcriptional level either leads to suppression of protein expression (orange) or increased turnover ratios and therefore noise reduction (blue).

predicted microRNA binding sites are indeed indicative of lower post-transcriptional production efficiencies. The data show that predicted microRNA targets have shorter mRNA half-lives and lower translation rate constants compared to matched non-targets; and differences increase for subsets of microRNA targets with higher number of binding sites for expressed microRNAs.

Next, in order to test how microRNA regulation is linked to transcriptional regulation, we matched control sets not only for 3'UTR length but also for equal post-transcriptional production efficiencies. We find that microRNA targets have higher transcription rates compared to control sets and that differences increase for subsets of microRNA targets with higher number of conserved binding sites (Figure 24A black). These results are thus consistent with the notion that microRNAs predominantly act to increase production ratios and therefore promote noise reduction in protein expression in mouse embryonic fibroblasts.

Equivalent analyses for genes harboring AU-rich sequences or Pumilio2 binding sites, motifs also associated with post-transcriptional repressive effects (Barreau et al., 2005, Zamore et al., 1997, White et al., 2001, Hafner et al., 2010), show that increased transcription rates for post-transcriptionally regulated genes is not the norm (Figure 24B+C).

Another set of data allowing us to investigate the coordination of microRNAs and transcriptional regulation comes from an experiment conducted by Baek et al. (2008). Here the effects of miR-223 in differentiation of mouse hematopoietic progenitors to

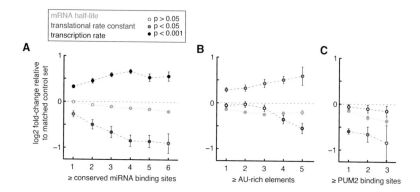

Figure 24: **MicroRNA targets in mouse embryonic fibroblasts have transcriptional and post-transcriptional rates consistent with increased production ratios**

(**A**) Differences in transcriptional and post-transcriptional rates between predicted microRNA targets and non-targets in mouse embryonic fibroblasts. For post-transcriptional parameters (mRNA half-life and translation rate constant) non-targets were matched for 3'UTR length. For transcription rates non-targets were matched for 3'UTR length as well as proteins per mRNA lifetime.

(**B**) and (**C**) same as in (A) but for genes with AU-rich elements and Pumilio2 binding sites, respectively.

In all panels error bars represent standard deviation of median ratios of target set to 1.000 control sets, p-values of equal medians compared to control sets calculated using two-sided Wilcoxon rank sum test.

neutrophil cells was investigated. miR-223 is already expressed in progenitor cells but up-regulated during the differentiation to neutrophil cells. Comparing mRNA expression levels before and after differentiation between wild-type and miR-223 deficient cells, the authors were able to disentangle mRNA expression changes during differentiation dependent and independent of miR-223. If microRNAs were participating in modulating production ratios rather than in suppressing unwanted protein expression, genes whose repression by miR-223 increases during differentiation should also show increased transcriptional activation. Indeed the data show that increases in miR-223-dependent repression during differentiation go hand in hand with miR-223-independent transcriptional up-regulation (Figure 25).

In summary, our analyses of the coordination of microRNA regulation and transcriptional regulation in mouse embryonic fibroblasts and during the differentiation

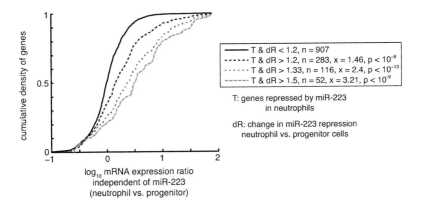

Figure 25: **Increased microRNA repression is linked to transcriptional upregulation during differentiation from mouse hematopoietic progenitor to neutrophil cells**
mRNA expression changes of miR-223 targets during differentiation from mouse hematopoietic progenitor to neutrophil cells (data from Baek et al. (2008)). Genes targeted by miR-223 in neutrophils that also increased in miR-223 repression during differentiation (gray shades, dashed lines) have progressively higher miR-223 *independent* mRNA expression (as determined from mRNA expression changes between miR-223 deficient progenitor and neutrophil cells) compared to neutrophil miR-223 targets whose miR-223 repression does not increase during differentiation (black). p-values were computed using two-sided Wilcoxon rank sum test.

of mouse hematopoietic progenitor to neutrophil cells presented here are consistent with the notion that microRNAs predominantly act to increase production ratios and thus reduce protein expression noise.

2.6   Noise-sensitive genes are enriched for microRNA regulation

To establish the role of microRNAs in regulating protein expression noise more firmly we set out to test whether classes of noise-sensitive genes are preferentially microRNA regulated.

We define genes that show altered functionality upon small perturbations of their expression as noise-sensitive. One such class are haploinsufficient genes, for which a 50% decrease in expression, due to disruption of one of their two alleles, is known to cause disease phenotypes. We reason that such genes should have mechanisms in place that limit their non-genetic expression fluctuations from deviating to levels that would cause such aberrant phenotypes in normal cells (Kemkemer et al., 2002, Magee et al., 2003).

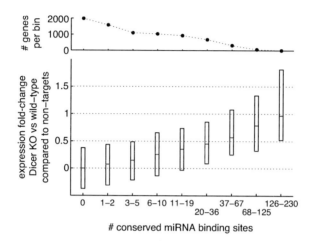

Figure 26: **Overall number of conserved microRNA binding sites per 3'UTR is a general marker for microRNA-mediated repression**
Fold-changes between Dicer knockout and wild-type mESC (data from Leung et al. (2011)) as a function of overall number of conserved microRNA binding sites in 3'UTRs. Genes are binned according to number of binding sites (division in log-space). Fold-changes are normalized to median fold-changes for non-targets (first bin). Upper panel shows number of genes per bin.

To inform about the extent of microRNA regulation of genes we will use the number of conserved microRNA target sites for all known microRNAs of an organism as a generic marker of microRNA-mediated repression across cell states. While it is clear that this simplification likely will not be valid for all genes or across all cell states, data from mouse embryonic stem cells support the notion that this marker is of general applicability to investigate microRNA-mediated repression organism-wide (Figure 26).

First we analyzed data from the Sanger Institute Mouse Genetics Project (White et al., 2013) on a diverse array of hetero- and homozygous knockouts in 6189 genes, which were tested for resulting aberrant phenotypes in mice. We classified all genes that show at least one phenotype in a heterozygous knockout (i.e. one of two alle-

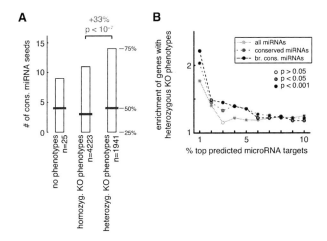

Figure 27: **Mouse noise-sensitive genes are enriched for overall microRNA regulation**

(**A**) Number of conserved microRNA binding sites for mouse genes classified according to their knockout phenotypes. Noise-sensitive genes, showing aberrant phenotypes already upon heterozygous knockouts, have more conserved microRNA binding sites than other genes, showing aberrant phenotypes only when both alleles are knocked out or not at all.

(**B**) Enrichment of genes with heterozygous KO phenotypes over homozygous KO phenotypes as a function of overall microRNA regulation. Top microRNA targets calculated from total number of conserved microRNA binding sites for all microRNAs (light gray), microRNAs conserved across most mammals (dark gray) or microRNAs conserved across most vertebrates (black) (classification from Friedman2009). p-values: Fisher's exact test.

les is dysfunctional) as 'noise-sensitive' (n=1941). Most other genes show at least one phenotype in homozygous knockouts (i.e. both alleles dysfunctional, n=4248). Only few genes show no phenotypes upon knockouts at all (n=25), which we therefore excluded from the analysis. We find that genes with heterozygous knockout phenotypes have a significant 33% enrichment in predicted microRNA binding sites compared to genes with homozygous knockout phenotypes (Figure 27A). When investigating the likelihood of phenotypes in heterozygous instead of homozygous knockouts as a function of microRNA regulation, we find that top predicted mouse microRNA targets are more likely to produce aberrant phenotypes upon heterozygous knockouts (Figure 27B). The 1% of mouse genes with the most conserved microRNA binding sites are twice as likely to show phenotypes upon heterozygous knockouts.

Next we turned to human noise-sensitive genes. We assembled three sets of noise-sensitive genes. Haploinsufficient genes, which we obtained from a literature survey (Dang et al., 2008), and genes implicated in the development of cancer. Here we distinguish two classes of noise-sensitive genes: Oncogenes, who cause susceptibility to cancer when their activity is increased, and tumor suppressors, who cause susceptibility to cancer when their activity is decreased (Hanahan and Weinberg, 2000). We assembled a list of 85 tumor suppressors and 331 oncogenes from the Cancer Gene Census (Bamford et al., 2004) and a recently published study (Davoli et al., 2013).

We find that haploinsufficient genes are more than three-fold enriched, tumor suppressors are more than six-fold enriched and oncogenes are more than four-fold enriched in predicted conserved microRNA binding sites compared to all other human genes (Figure 28A). Similar to mouse noise-sensitive genes we find that the likelihood to be noise-sensitive (i.e. belonging to either one of the three categories of human noise-sensitive genes) increases for the top human microRNA targets (Figure 28B). The likelihood to be noise-sensitive is up to five-fold higher for the top 1% of human microRNA targets.

It is likely that not all genes from the literature-curated lists are equally noise-sensitive. We sought to improve our confidence in the noise-sensitivity of tumor suppressors and oncogenes by making use of the enrichment of genes for certain types of alterations found in cancer sequencing data, as calculated by Davoli et al. (2013). We reasoned that genes enriched for certain types of alterations that are clearly linked to expression changes should indicate increased noise-sensitivity. 56, out of the 85 literature-defined, tumor suppressors were found to be significantly enriched for loss-of-function mutations. We find these tumor suppressors to have three times as many conserved microRNA binding sites compared to the remaining 29 tumor suppressors, which were not enriched for loss-of-function mutations (Figure 28C). We find similar results, albeit with smaller differences, for tumor suppressors enriched for copy number loss and oncogenes enriched for copy number amplifications. Of note, we find oncogenes that are enriched for recurring, localized mutations in their coding sequence to be depleted for microRNA binding sites compared to all other oncogenes. This subset of oncogenes is enriched for genes acting in signaling pathways that are often robust to expression changes (e.g. mediated through feed-

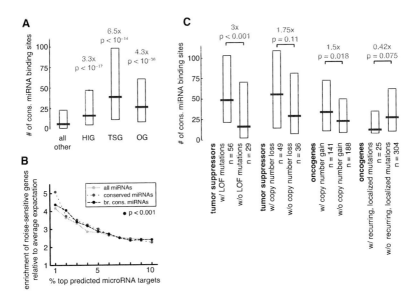

Figure 28: **Human noise-sensitive genes are enriched for conserved microRNA binding sites**

(**A**) Number of conserved microRNA binding sites for haploinsufficient genes (HIG), tumor suppressors genes (TSG) and oncogenes (OG) compared to all other genes. p-values for equal medians of noise-sensitive sets and background set calculated using two-sided Wilcoxon rank sum test.

(**B**) Enrichment of noise-sensitive genes as function of conserved microRNA binding sites in 3'UTRs. Each data point shows the probability of the X% of genes with the most conserved microRNA binding sites to be noise-sensitive relative to the average expectation. p-values for deviation in the number of noise-sensitive genes calculated using Fisher's exact test.

(**C**) Tumor suppressors and oncogenes were further distinguished by their enrichment for loss-of-function (LOF) mutations, genomic copy number loss (both TSG), genomic copy number amplification or recurring localized mutations in cancer sequencing data (both OG). p-value for equal medians of both sets calculated using two-sided Wilcoxon rank sum test.

backs) and instead achieve full oncogenic potential only through specific mutations that alter their protein activity (Legewie et al., 2009, Fritsche-Guenther et al., 2011).

We further attempted to control for confounding factors that could potentially lead to noise-sensitive genes being enriched for microRNA regulation.

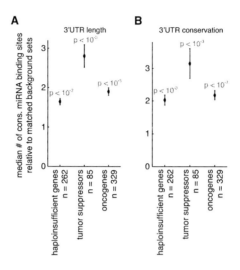

Figure 29: **Controlling for 3'UTR length and conservation**
(**A**) Median enrichment of conserved microRNA binding sites in noise-sensitive sets relative to 1000 3'UTR length matched control sets.
(**B**) Median enrichment of conserved microRNA binding sites in noise-sensitive sets relative to 1000 3'UTR conservation matched control sets.
Indicated p-values are mean of log-transformed p-values for equal medians of noise-sensitive sets and each control set calculated using two-sided Wilcoxon rank sum test.

First, the enrichment for conserved microRNA binding sites could be affected by the length and conservation of 3'UTRs. Controlling for 3'UTR length shows that noise-sensitive genes are enriched for conserved microRNA binding sites compared to genes with similar 3'UTR lengths (Figure 29A). Next we computed the average base-wise conservation of 3'UTRs using PhyloP conservation scores, which were calculated from multi-sequence alignments of the human genome to 45 vertebrate genomes (Pollard et al., 2010). We find that noise-sensitive genes are enriched for conserved microRNA binding sites compared to genes with similar 3'UTR conservation (Figure 29B).

Second, the tissue-specific expression of both genes and microRNAs could bias the number of microRNA binding sites for subsets of genes. We first analyzed how ubiquitously genes are expressed across tissues. We obtained mRNA expression

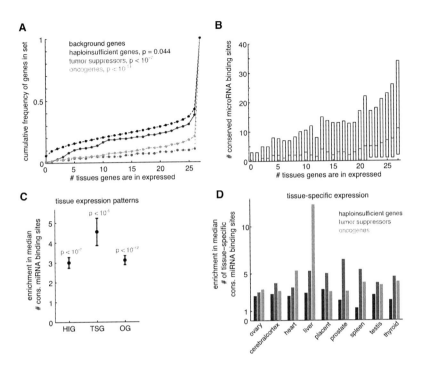

Figure 30: **Controlling for tissue-specific expression patterns**

(**A**) Cumulative frequency of genes as a function of the number of tissues they are expressed in. p-values for equal distributions of noise-sensitive sets compared to background set calculated using Kolmogorow-Smirnow test.

(**B**) Number of conserved microRNA binding sites as a function of how many tissues genes are expressed in.

(**C**) Median number of conserved microRNA binding sites of noise-sensitive sets relative to 1000 background sets matched for the number of tissues the genes are expressed in. Indicated p-values are mean of log-transformed p-values for equal medians of noise-sensitive sets and each background set calculated using two-sided Wilcoxon rank sum test.

(**D**) Median number of tissue-specific conserved microRNA binding sites of noise-sensitive sets relative to 1000 background sets matched for the tissue-specific expression. All enrichments significant at $\overline{log_{10}(p_i)} < -3$, with p_i as p-value for equal medians of noise-sensitive sets and each background set calculated using two-sided Wilcoxon rank sum test.

levels of genes across 27 human tissues (Fagerberg et al., 2014). The data show that about 60% of genes are expressed ubiquitously, while the remaining 40% are expressed in a more tissue-specific manner (Figure 30A). Noise-sensitive genes show a significant tendency to be expressed more ubiquitously across tissues compared to other genes.

Because microRNAs, too, are expressed in a tissue-specific manner, genes expressed in more tissues tend to have higher number of microRNA binding sites (Figure 30B). To ensure that higher number of microRNA binding sites of noise-sensitive genes are not solely a consequence of their ubiquitous expression across tissues, we compared them to control sets of genes matched for the number of tissues they are expressed in (Figure 30C). Consistent with the enrichment not being due to tissue-expression patterns, we find that the enrichment of microRNA binding sites on noise-sensitive genes compared to tissue-expression matched control sets is nearly as large as when compared against all human genes (Figure 28A).

Furthermore, tissue-specific microRNA regulation can also depend on the expression levels of genes (Figure 21; and (Farh et al., 2005, Stark et al., 2005, Sood et al., 2006)), where lowly expressed genes are often enriched for binding sites of tissue-expressed microRNAs, while highly expressed genes selectively avoid those binding sites (Farh et al., 2005, Sood et al., 2006). To control for this effect we obtained tissue-specific microRNA expression patterns from the human microRNA expression atlas (Landgraf et al., 2007). Combining this information with the data on tissue-specific mRNA expression yielded nine human tissues with data for the expression of microRNAs as well as genes. For each set of noise-sensitive genes we created tissue-specific control sets of genes with equal expression distributions. We find that noise-sensitive genes are significantly enriched for tissue-specific microRNA binding sites compared to similarly expressed genes in all tested tissues (Figure 30D). However, there is considerable variation in the enrichment for sets across the tested tissues.

The expression-dependent phenotypes, which we have based our noise-sensitivity classification on, are likely to some degree tissue-specific (Davoli et al., 2013). It is therefore important to understand whether noise-sensitive genes are actually micro-RNA-regulated in those circumstances where genes exhibit their noise-sensitivity. We used a list of predicted cancer-specific tumor suppressors (Davoli et al., 2013) to test whether they are indeed enriched for binding sites of microRNAs specific to these cancers or their tissues of origin. To do so we assembled microRNA expression profiles from the human microRNA expression atlas (Landgraf et al., 2007)

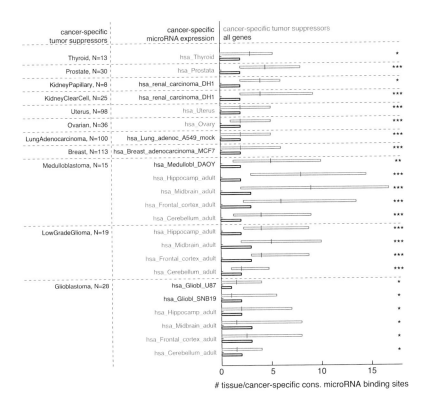

Figure 31: **Cancer-specific tumor suppressors are enriched for conserved binding sites of microRNAs specific to respective cancers and related tissues**
Number of conserved binding sites for microRNAs expressed in cancers or tissues in which genes are predicted to act as tumor suppressors. Left column gives cancer types and number of predicted cancer-specific tumor suppressors (Davoli et al., 2013). Middle column gives sample name of matched cancer (black) or tissue (gray) microRNA expression (Landgraf et al., 2007). Data: distributions for cancer-specific tumor suppressors in gray, for all other genes in black. p-values for equal medians calculated using two-sided Wilcoxon rank sum test. ***: $p < 10^{-5}$, **: $p < 10^{-3}$, *: $p < 0.05$

matching those cancers for which cancer-specific tumor suppressors were predicted. We find that predicted cancer-specific tumor suppressors are significantly enriched in conserved binding sites for those microRNAs specific to the cancers or their tissues of origin (Figure 31). This indicates that microRNA regulation of tumor suppressors is indeed in place where precision of expression matters.

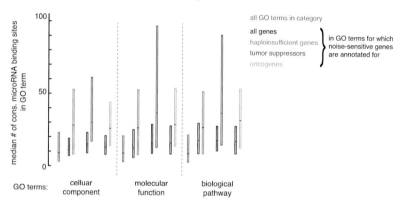

Figure 32: **Noise-sensitive genes are enriched for microRNA binding sites compared to genes in the same cellular components, with the same molecular functions or active in the same biological pathways**
Shown are the distributions of median number of conserved microRNA binding sites of genes in Gene Ontology terms. Terms are stratified in categories according to cellular components, molecular functions and biological pathways. Grey bars show median microRNA targeting of genes in all GO terms of the category (left most in category). Black bars show median microRNA targeting of all genes in GO terms for which the respective noise-sensitive genes are annotated (left of respective noise-sensitive class). Colored bars show median microRNA targeting of noise-sensitive genes in terms for which they are are annotated (green, haploinsufficient; blue, tumor suppressors; yellow, oncogenes). Noise-sensitive genes are significantly enriched in microRNA targeting over other genes annotated for same terms ($p < 10^{-14}$, paired Student's t-test). GO terms for which noise-sensitive genes are annotated are significantly enriched in microRNA targeting over all GO terms in category ($p < 10^{-6}$ for all comparisons, except GO terms of cellular components for which haploinsufficient genes are annotated ($p < 0.001$), two-sided Wilcoxon rank sum test).

The expression-dependent phenotypes of noise-sensitive genes are likely mediated through the cellular processes they act in. We were therefore interested to understand whether noise-sensitive genes are also enriched in microRNA regulation compared to genes in acting in the same cellular processes or whether these processes are generally enriched for microRNA regulation. To inform about the cellular processes that noise-sensitive genes act in we assembled Gene Ontology (GO) terms for cellular components, molecular function and biological pathways of all human genes (Ashburner et al., 2000).

We first investigated microRNA regulation in all GO terms for which at least one gene from the noise-sensitive sets is annotated. We calculated the median number of conserved microRNA binding sites in each of these GO terms for (a) all noise-sensitive genes annotated for the GO term and (b) all other, non-noise-sensitive genes annotated for the GO term. We find that noise-sensitive genes are significantly enriched in microRNA regulation compared to other genes annotated for the same GO terms (Figure 32, colored distributions).

However, when comparing the median microRNA regulation of all genes annotated in 'noise-sensitive' GO terms to the median microRNA regulation of all GO terms in the same category, we find that 'noise-sensitive' GO terms are significantly enriched for microRNA regulation (Figure 32, black versus grey distributions). This suggests that the overall enrichment of microRNA regulation in a GO term could be a proxy for the noise-sensitivity of the annotated cellular process itself. We therefore calculated the median number of binding sites in genes across all GO terms and determined GO terms enriched or depleted for microRNA regulation (see section 4.4 for the complete lists). Unsurprisingly, GO terms enriched or depleted for microRNA regulation are similar to those GO terms that are enriched for genes with long and short 3'UTRs (Ramsköld et al., 2009).

We find that GO terms enriched for microRNA regulation can be roughly divided into three major groups. First, GO terms associated with regulation on the transcriptional and post-transcriptional levels and macromolecular complexes associated therewith. Second, GO terms participating in organismal development, particularly that of the nervous system, with a subgroup being associated with numerous cellular neuronal components. Third, GO terms participating in signaling pathways, including in particular terms for protein-protein interactions and protein modifications.

GO terms depleted for microRNA regulation can be roughly divided into four major groups. First, GO terms associated with the translational process (especially ribosomal components) and protein degradation. Second, GO terms associated with the immune system and extracellular components, including secretion. Third, GO terms associated with metabolism and the mitochondria. Four, GO terms related to olfactory and other sensory functions.

These results are very intriguing. GO terms enriched for microRNA regulation are involved in regulatory and developmental cellular processes. They would thus be expected to require genes with low protein noise in order to maximize accuracy but also to minimize the propagation of noise to other genes and processes. On

59

the contrary, GO terms depleted for microRNA regulation either describe constitutive functions that likely involve large number of molecules (protein production and degradation, metabolism) or processes performed by large cell populations (extracellular components/secretion, immune system). They thus either might not need tight control of noise or low noise level might be achieved instead by averaging over many molecules or participating cells. Furthermore, olfactory neurons have been proposed to deploy stochastic gene expression of olfactory receptors and negative feedbacks to ensure mono-allelic production of one specific olfactory receptor per neuron (Chess et al., 1994, Serizawa et al., 2004).

Both, GO terms enriched as well as depleted for microRNA regulation, therefore fit well with what we would expect if the function of microRNA regulation were to decrease protein expression noise.

In summary, we find that mouse and human noise-sensitive genes are enriched for microRNA binding sites, suggesting a role for microRNAs to minimize noise in the production of these genes. This is corroborated by our findings that genes implicated in cancer that show strong enrichment for expression-altering aberrations are further enriched for microRNA binding sites, while those oncogenes that are likely buffered against the effects of altered expression levels in normal cells are depleted for microRNA binding sites. Furthermore, predicted cancer-specific tumor suppressors are enriched for binding sites of microRNAs specific to their cancers or the respective tissues of origin, suggesting that microRNAs indeed act where noise reduction should be necessary. The finding that the biological processes that noise-sensitive genes act in are themselves enriched for microRNA regulation suggests that the overall microRNA regulation of a biological processes can be a proxy for its noise-sensitivity. Indeed, the patterns of biological processes enriched and depleted for microRNA regulation fit well with the general expectation for noise-sensitivity.

3 Conclusions and Outlook

3.1 Protein expression noise in mammalian cells

In Chapter 1 I have investigated whether genes in mammalian cells control intrinsic noise in protein expression noise via their production ratios. Intrinsic noise in protein expression generally declines with increasing expression but is also influenced by the ratio of contributions from the transcriptional and post-transcriptional level to protein production (Thattai and van Oudenaarden, 2001, Elowitz et al., 2002, Ozbudak et al., 2002, Paulsson, 2004, Pedraza and Paulsson, 2008). My analysis of expression parameters in mouse embryonic fibroblasts showed that genes exhibit as much as two orders of magnitude differences in their production ratios (transcription rate divided by post-transcriptional efficiencies) to translate similar amounts of protein. Here, genes more prone to noise - because of low expression or short-lived proteins -, and those with noise-sensitive functions (transcription factors), exhibit production ratios that correspond to low noise arising in the production of their proteins. This therefore indicates that genes in mammalian cells control intrinsic noise in protein expression noise via their production ratios, in line with results from bacteria (Ozbudak et al., 2002) and yeast (Fraser et al., 2004).

The balance of intrinsic and extrinsic noise

Minimization of intrinsic noise in protein expression via large production ratios is only reasonable when protein expression is dominated by intrinsic noise. Proteins dominated by extrinsic noise should instead profit little from minimizing intrinsic noise. The data presented in chapters 2.3 and 2.4 show that for the tested mCherry reporter intrinsic noise dominates over the majority of the expression range of endogenous genes in mouse embryonic stem cells, suggesting that protein expression noise for most genes can be substantially altered via production ratios. Of note, studies conducted in *E.coli* and yeast that estimated intrinsic and extrinsic noise in numerous endogenous genes via fluorescent reporter knock-in libraries have arrived at similar conclusions about the balance of intrinsic and extrinsic noise (Taniguchi et al., 2010, Stewart-Ornstein et al., 2012). However, as mentioned in chapter 2.4, the balance of intrinsic and extrinsic noise might vary for genes with expression parameters different from those of our reporter setup. This also extends to cell types other than mouse embryonic stem cells, where cell type-specific parameters, such as cellular volumes, might shift the balance between intrinsic and extrinsic noise. Critical assessment of the actual noise balances of genes in mammalian cells must therefore come from future experiments.

Here two approaches come to mind. A gold standard approach to determine the noise balances would be labeling both alleles of genes with different fluorescence reporters via genetic knock-ins. The intracellular allelic expression differences would then inform about intrinsic noise (similar to the intrinsic noise experiments presented in Figure 13) and the intercellular expression differences would inform about total noise, from which extrinsic noise could be deduced. Such an approach seems feasible with modern genome editing approaches, such as the CRISPR system (Jinek et al., 2012, Cong et al., 2013), for select genes, but assaying a wide array of genes to arrive at a comprehensive picture of noise balances in protein expression would be laborious.

The recent advent of techniques that can measure single cell transcriptomes holds the promise for an orthogonal approach, allowing to measure the expression of numerous genes in many individual cells at once with relative ease (Hashimshony et al., 2012, Grün et al., 2014, Islam et al., 2014, Deng et al., 2014, Buettner et al., 2015). Investigating the transcriptomes of chimeric cells (with both genome copies coming from different backgrounds, therefore harboring numerous single nucleotide mutations to distinguish them) could in principle be used to elucidate the balance of intrinsic and extrinsic noise Deng et al. (2014). However, such measurements would need to be combined with complementary measurement of mRNA and protein half-lives as well as translation rates to yield meaningful estimates of noise on the protein level. This is because the relation between mRNA and protein half-lives determines how much noise propagates from the mRNA to the protein level and because effects of post-transcriptional regulation, which we have shown to be of importance in noise control, only become apparent through the post-transcriptional rates.

3.2 microRNA control of protein expression noise in mammalian cells

If noise control is a common theme in the production of genes, who are the regulators that mediate it? And what can they tell us about the extent of noise control in mammalian organisms?

In chapters 2.2 and 2.3 I conducted a theoretical and experimental analysis about the potential of microRNAs to control protein expression noise. I used a mathematical model to generate predictions about microRNA-mediated noise effects and a synthetic plasmid reporter system to test their validity in mouse embryonic stem cells. The analyses showed that microRNAs repress intrinsic noise in protein production as a function of how strongly they repress a gene, which is equivalent to the shifting of production ratios as discussed in chapter 1. On the contrary, noise in the pool of microRNAs propagates to the regulated gene, leading to a net increase of noise at high expression levels, where intrinsic noise is diminished. However, combinatorial targeting, commonly observed for endogenous microRNA targets (Enright et al., 2003, Stark et al., 2003, John et al., 2004, Krek et al., 2005), can mitigate this additional noise by averaging over fluctuations in independent microRNAs pools. In chapter 2.4 I have explored how the observed microRNA-mediated noise effects translate to endogenous target situations. Native 3'UTRs of endogenous microRNA targets show pronounced noise reduction over wide expression ranges likely due to combinatorial regulation, which results in strong repression and low noise in mixed pools. The expression range of noise reduction reaches beyond the 90th percentile of the mESC transcriptome and a comparison of transcriptome expression in wild-type and microRNA-deficient mESC showed that hundred of genes within this expression range are strongly repressed by mESC microRNAs, suggesting they should have reduced protein expression noise due to their microRNA regulation.

The apparent potential of microRNA regulation to decrease protein expression noise raises the question whether noise reduction is a biologically significant function of microRNA regulation or a mere side-effect of repressing protein expression with little physiological significance.

The 'purpose' of microRNA regulation

Two major ideas for functions of microRNAs have been proposed: suppression and tuning of protein expression (Reinhart et al., 2000, Bartel and Chen, 2004). In chapter 2.5 I have argued that these two functions should crucially differ in how they are coordinated with transcriptional regulation. Suppression should be linked with repressed transcriptional activity, while tuning should be linked to increased transcriptional activity. I have then shown that in mouse embryonic fibroblasts and in a differentiation assay of mouse hematopoietic cells, (increased) microRNA regulation is on average associated with higher transcription rates, therefore arguing for a prevalent tuning function of microRNAs in these two systems. Furthermore, the findings that noise-sensitive genes in mouse and human are associated with a strong enrichment for microRNA regulation further argues against a suppressive function of microRNAs.

The observation that most strongly repressed microRNA-targets in mouse embryonic stem cells are lowly expressed is consistent with previous studies, that have shown predicted tissue-specific microRNA binding sites to be enriched for lowly expressed genes, while highly expressed genes are selectively depleted for microRNA binding sites (Farh et al., 2005, Stark et al., 2005, Sood et al., 2006). These results were interpreted with a suppression function of microRNAs in mind, assuming that microRNA targets are lowly expressed because of microRNA regulation and that highly expressed genes avoid microRNA regulation in order to achieve their high expression levels. The expression-dependent characteristics of microRNA-mediated noise effects lend an alternative explanation to these observations: Lowly expressed, noise-prone genes might be under evolutionary pressure to acquire microRNA-mediated noise reduction, while highly expressed genes are less noise-prone and would instead suffer increased noise as a consequence of microRNA regulation that might therefore be selected against. Furthermore, the aforementioned studies reported that targets of tissue-specific microRNAs are often those genes that are expressed lower in the specific tissues than they are elsewhere (Farh et al., 2005, Sood et al., 2006). This observation is also compatible with a noise-centric perspective. Ubiquitously expressed genes might have constitutive post-transcriptional regulation adjusting expression noise to their organism-wide needs (see below). However, in tissues where a lower expression is optimal (making them more noise-prone) or where their functions are more sensitive to noise they would be under selective pressure to acquire additional tissue-specific microRNA regulation to reduce noise.

The possibility that a primary function of microRNA regulation is noise reduction demands a critical evaluation of how physiological roles are assigned to specific microRNAs. For example, if a microRNA regulates a gene whose expression is needed to induce a cell cycle transition, does this microRNA then negatively regulate the transition by repressing the gene or does it instead promote the transition by ensuring precise expression distributions of the gene that therefore enable the transition? Discriminating these two possibilities might be very difficult due to the entanglement of mean and noise levels. To test whether noise levels are important, noise levels would have to be varied by altering microRNA regulation while simultaneously keeping mean expression levels in check by adjusting transcription rates accordingly. However, this might be difficult especially in processes such as the cell cycle, where regulators presumably are not expressed at constant mean levels throughout the whole cycle.

The general potential of post-transcriptional processes for noise control

In the second part of my thesis I have focused on the role of microRNAs, arguably the most prominent post-transcriptional regulators in multicellular organisms, in regulating production ratios. However, it is clear that production ratios, and therefore noise, can be regulated post-transcriptionally in various ways ((Ozbudak et al., 2002, Raser and O'Shea, 2005), also exemplified by the reporter experiments with AU-rich elements presented in Figure 15). On the post-transcriptional level two classes of regulation can be distinguished. Intrinsic features of the mRNA, such as the sequence and secondary structure around the translation initiation site (Bentele et al., 2013, Noderer et al., 2014) or the (initial) length of the polyA-tail (Subtelny et al., 2014), that directly affect post-transcriptional rates; and extrinsic regulators, such as microRNAs and other RNA-binding proteins (Baltz et al., 2012, Castello et al., 2012), that bind to sequence or structure elements in the mRNA to regulate post-transcriptional rates (Goodarzi et al., 2012, Oikonomou et al., 2014). These two classes of regulation have different advantages. Intrinsic features will regulate post-transcriptional rates without additional extrinsic noise propagating to the protein. On the contrary, while extrinsic regulators always come at the cost of additional extrinsic noise, they can act in cell type-specific ways by means of altered expression of the regulator or alternative 3'UTR isoforms (regulated via alternative cleavage and polyadenylation (Edwalds-Gilbert et al., 1997)). I would therefore expect that genes expressed at similar levels and with similar needs for noise control across cells of the entire organism should mainly be regulated by intrinsic features, as long as these can achieve the whole breath of post-transcriptional efficiencies necessary. However the pervasive post-transcriptional regulation of genes by external regulators (Enright et al., 2003, Lewis et al., 2003, Stark et al., 2003, Baltz et al., 2012, Castello et al., 2012) as well as the extensive use of differential 3'UTR isoforms (Sandberg et al., 2008, Mayr and Bartel, 2009, Li et al., 2012, Ulitsky et al., 2012) indicates that many genes have the need for cell type-specific adjustment of their post-transcriptional efficiencies. Intriguingly, increasingly longer 3'UTR isoforms are found in the progression through zebrafish development, with the longest 3'UTR isoforms found in brain (Li et al., 2012, Ulitsky et al., 2012). An increase in post-transcriptional regulation by microRNAs and other external regulators by means of longer 3'UTR isoforms throughout development is consistent with our analysis of Gene Ontology terms enriched for microRNA regulation in human (presented in chapter 2.6). This therefore corroborates the notion that microRNA regulation lends increasing precision to the gene expression programs of differentiating cells and can thereby stabilizes ever more specialized cell states.

In conclusions, in this thesis I have presented evidence that protein expression noise is a relevant parameter of genes in mammalian cells that is in part controlled by production ratios. I have shown that microRNA regulation - in conjunction with increased transcription - can effectively shift these production ratios to reduce noise. The characteristics of endogenous microRNA-target interactions - combinatorial regulation and targeting of lowly expressed genes - are suggestive of microRNA regulation being set up to reduce protein expression noise. Evidence that microRNA regulation is indeed deployed for this purpose comes from further findings that microRNA regulation goes hand in hand with increased transcription of target genes and that noise-sensitive genes as well as regulatory and organizing cellular and developmental processes are enriched for microRNA regulation.

4 Materials & Methods

4.1 Bioinformatics

Reference genomes

All computations for mouse were performed on RefSeq gene annotations build GRCm38/mm10 (Dec. 2011). All computations for human were performed on RefSeq gene annotations build GRCh37/hg19 (Feb. 2009). Annotations were downloaded from the UCSC genome browser (http://genome.ucsc.edu/) with entries for chromosome, strand, transcription start and end positions, coding sequence start and end positions, exon start and end positions, associated UniProtID, GeneSymbol and RefSeq IDs.

Predicted regulation by microRNAs, AU-rich elements and Pumilio 2

Data on microRNA binding sites in the 3'UTRs of mouse and human genes were downloaded from Targetscan (http://www.targetscan.org, version 6.2, June 2012, (Garcia et al., 2011)). For all analyses (except as presented in Figure 29C) we only considered conserved microRNA binding sites. AU-rich elements ('ATTTA', (Barreau et al., 2005)) and Pumilio 2 binding sites ('TGTA[A|T|G|C]ATA', (Hafner et al., 2010)) in 3'UTRs of mouse genes were determined using a custom Matlab script.

Expression parameters in mouse embryonic fibroblasts

Expression parameters were obtained from supplementary data of Schwanhäusser et al. (2011). Entries were matched to RefSeq gene annotations. Entries for which not all expression parameters were estimated in both replicates were discarded. Also entries for which the standard deviation of protein expression levels between replicates was larger than the mean of both replicates were discarded. This resulted in expression parameters for n = 2815 genes. Genes were divided into three bins with equal number of genes according to their translation rates. For analyses of transcriptional and post-transcriptional parameters as a function of protein half-lives (Figure 3) genes in each translation rate bin were divided into two subsets according to their protein half-lives. Translation rate distributions in both subsets were matched. This was done by subdividing the log-transformed translation rate distribution into ten equally spaced bins and matching number of genes per subset in each bin by randomly discarding surplus genes from the subset with more genes in the bin. Equivalency of resulting translation rate distributions of subsets was ensured with two-sided Wilcoxon rank sum test. Medians of transcriptional and post-transcriptional parameters were compared between both subsets. This procedure was repeated 1,000 times. The means and standard deviations of median ratios over all 1,000 iterations are reported.

For the analysis of transcriptional and post-transcriptional parameters as a function of microRNA binding sites, AU-rich elements, or Pumilio 2 binding sites (Figure 24) targetsets of genes with more than a respective number of binding sites were defined. For microRNA binding sites we only considered those of microRNAs expressed in mouse embryonic fibroblasts (data from mouse microRNA expression atlas (Landgraf et al., 2007)). For analysis of post-transcriptional rates for each gene in the respective targetset the nearest ten neighbors in the 3'UTR length space were chosen. Only genes with 3'UTRs longer 100 nucleotides were considered. Similar to the analysis of parameters as a function of protein half-lives, 3'UTR length distributions were matched between targetset and nearest neighbor set and the post-transcriptional rates of the resulting sets compared in 1,000 iterations. To analyze transcription rates we additionally matched targetset and nearest neighbor set for post-transcriptional efficiencies (number of proteins produced per mRNA lifetime). This was accomplished by subdividing the joint log-transformed distributions of 3'UTR length and post-transcriptional efficiencies into 10x10 bins. In each of the one hundred bins, number of genes in both sets were matched and the transcription rate of the resulting sets were compared in 1,000 iterations.

microRNA regulation during differentiation of mouse hematopoietic cells

Differential expression data of mRNA and proteins levels between wild-type and miR-223 deficient hematopoietic progenitor (only mRNA) and neutrophil cells was obtained from supplementary data of Baek et al. (2008). We only considered genes that a) are repressed on the mRNA level by miR-223 in neutrophil cells (as determined by fold-changes of mRNA levels between wild-type and miR-223 deficient neutrophil cells); b) were measured on the mRNA level in all states (wild-type and miR-223 deficient progenitor and neutrophil cells); c) were measured on the protein level in neutrophil cells in at least six independent measurements (to ensure genes are robustly expressed). miR-223 repression fold-changes during differentiation were calculated from differences in fold-changes in wild-type and miR-223 deficient progenitor and neutrophil cells. mRNA fold-changes independent of miR-223 are reported by Baek2008 as fold-changes in mRNA levels between miR-223 deficient progenitor and neutrophil cells.

Conserved microRNA binding sites as proxy for microRNA repression

Microarray expression data from wild-type and Dicer knockout mouse embryonic stem cells Leung et al. (2011) were obtained from Gene Expression Omnibus GSE25310. MicroRNA-mediated repression was calculated as the fold-change in mean expression between wild-type and Dicer knock-out samples. Only genes above mean microarray intensity of $log_2(5)$ in wild-type samples were considered. Distribution of gene-wise number of conserved microRNA binding sites was log-transformed (a +1 pseudo-count was added) and divided into nine equally spaced bins. Bin-wise distributions of microRNA-mediated repression were normalized by median microRNA-mediated repression in first bin (with genes having no microRNA binding sites).

Analysis of mouse noise-sensitive genes

A table containing phenotype-genotype relationships ('MGI_PhenotypicAllele.rpt') as well as a table describing the mammalian phenotype annotations ('MGI_PhenoGenoMP.rpt') was downloaded from the Mouse Genome Database consortium (http://www.informatics.jax.org/, November 2013, (Eppig et al., 2015)). From phenotype-genotype relationship table GeneIDs of knocked-out genes, the zygosity of the specific knockout and the observed phenotypes were extracted. Mammalian phenotype annotations were used to classify phenotypes. All phenotypes were considered aberrant except those that fell into the two categories 'normal phenotype' and 'no phenotypic analysis'. Genes were classified as 'heterozygous knockout phenotypes' when at least one of their reported heterozygous knockouts showed an aberrant phenotype. Genes were classified as 'homozygous knockout phenotypes' when none of their reported heterozygous knockouts showed any phenotype but at least one of their reported homozygous knockouts did.

Analysis of human noise-sensitive genes

Human haploinsufficient genes were obtained from supplementary data of Dang et al. (2008). Human tumor suppressors (TSGs) and oncogenes (OGs) were as-

sembled from the Cancer Gene Census
(`http://cancer.sanger.ac.uk/cancergenome/projects/census/`,
(Futreal et al., 2004)). TSGs were defined as those genes acting in a recessive manner and have reported somatic mutations. OGs were defined as those genes acting in a dominant manner with somatic mutations. Lists were further supplemented with 14 TSGs and four OGs reported by Davoli et al. (2013). Enrichment for alterations (loss-of-function mutations, genomic deletions, genomic amplifications, recurrent localized mutations ['Entropy Missense mutations']) in cancer sequencing data were obtained from supplementary data of Davoli et al. (2013). TSGs and OGs were reported as enriched for these alterations at q-values below 0.05. Predicted cancer-specific TSGs were also obtained from the supplementary data of Davoli et al. (2013). MicroRNA expression profiles of cancers and related tissues were obtained from the human microRNA expression atlas (Landgraf et al., 2007) and matched as indicated in Figure 31.

Gene Ontology analyses

The Ontology (24/03/2014) and GO annotations for human (24/03/2014) and mouse (16/11/2014) were downloaded from `http://geneontology.org/`. GO annotations from the three categories ('cellular component', 'molecular function', and 'biological pathway') were mapped onto RefSeq genes using the Matlab function `goannotread` and a custom script. For the analysis of production ratios of transcription factors in mouse embryonic stem cells (Figure 4) we extracted genes annotated for GOid 3700 (molecular function "sequence-specific DNA binding transcription factor activity").

Enrichment or depletion of GO terms for human conserved microRNA binding sites was calculated by comparing the distribution of microRNA binding sites of genes annotated for a specific GO term to the distribution of number of microRNA binding sites of all genes annotated for GO terms in the category (here genes were listed as many times as they are annotated for GO terms in the category). p-values for enrichment or depletion were computed using one-sided Wilcoxon rank sum test, multiple testing corrected with `mafdr` algorithm in Matlab, and reported in Section 4.4 if smaller than 0.01.

microRNA targeting matched for 3'UTR length and conservation

Base-wise conservation scores (PhyloP46) of 45 vertebrate genomes with human genome (hg19) were downloaded from `http://genome.ucsc.edu/` (Pollard et al., 2010). Mean of base-wise conservation scores for each 3'UTR was calculated. Analyses of microRNA regulation of noise-sensitive genes compared to 3'UTR length or conservation matched background sets was performed as described above in section 'Expression parameters in mouse embryonic fibroblasts'.

Human tissue expression data

Human transcriptome data across 27 tissues (Fagerberg et al., 2014) were downloaded from ArrayExpress (`https://www.ebi.ac.uk/arrayexpress/experiments/E-MTAB-1733/`, (Kolesnikov et al., 2015)). FPKM expression levels from multiple samples of same tissues were averaged. Genes with non-zero FPKM values in tissue were counted as expressed (conclusions do not change for higher cutoffs). To control microRNA regulation of noise-sensitive genes for tissue expression patterns a set of genes from background (all non noise-sensitive genes) were randomly picked to match the distribution of number of tissues noise-sensitive genes are expressed in. Medians of number of conserved microRNA binding sites were compared. This was repeated 1,000 times. Reported are means and standard deviation of median ratios over all iterations.

To control microRNA regulation of noise-sensitive genes for tissue-specific expression levels, tissues transcriptome data were matched with corresponding tissue microRNA expression data from human microRNA expression atlas (Landgraf et al., 2007). Background sets of genes with same mRNA expression distribution as noise-sensitive genes were randomly chosen with nearest neighbor approach as described in section 'Expression parameters in mouse embryonic fibroblasts'. Median number of conserved binding sites for microRNAs expressed in the tissue were compared between noise-sensitive gene set and background set. This was repeated 1,000 times. Mean of median ratios is reported.

4.2 Experiments

Reporter plasmid construction

Starting from a previously established reporter system (Mukherji et al., 2011), eYFP was replaced with ZsGreen1-1 (Clontech) using EcoRI and NdeI digestion sites. MicroRNA binding sites were inserted into mCherry 3'UTR using ClaI and EcoRV digestions sites and into ZsGreen1-1 3'UTR using NdeI and XbaI digestion sites. N = 1 bulged (full complementary to microRNA except central bulge, as in Mukherji et al. (2011)) and perfect (full complementary) microRNA target sites were created by aligning complementary single stranded oligonucleotides with respective over-hangs for digestions sites (IDT) at 65°C for 30 minutes, with previous heating to 95°C for 5 minutes. N = 4 bulged miR-20a binding site 3'UTR contains random 50bp spacers between individual binding sites and was synthesized (IDT gBlocks). Wee1 wild-type and mutated 3'UTR fragments (positions 130-610) as well as Casp2 and Rbl2 wild-type and mutated 3'UTRs were synthesized (IDT gBlocks). The Lats2 wild-type 3'UTR was amplified from murine embryonic stem cell cDNA (forward primer AATAAGGATCCCGAGGAAACCCAAAATGAGATTTCTTTTC, reverse primer AACAAGCTAGCGGCTTTAAAGTTTTAATAATAAATTGTGCCAGTAGA) and was se-quence confirmed. The mutated version of Lats2 3'UTR was synthesized (Gen-eArt). The mutated 3'UTRs were synthesized with double point mutations in all predicted microRNA binding sites (Targetscan6.2 (Garcia et al., 2011)) of signifi-cantly expressed mESC microRNAs (Marson et al., 2008). Seed positions 3 and 5 were mutated such that purines and pyrimidines were interchanged, yielding mu-tated 3'UTRs that maintain >95% sequence similarity to wild-type 3'UTRs. Synthe-sized fragments were PCR amplified to append necessary digestion sites. MicroRNA binding sites and 3'UTRs were cloned into digested and dephosphorylated plasmid backbone using T4 ligase (NEB). For a list of target site sequences, endogenous 3'UTR sequences and their mutated versions refer to Section **??**.

Transient transfections

Murine embryonic stem cells V19 below passage 20 were plated two days before transfection in 2 ml synthetic 2i medium (Ying et al., 2003) (Gibco) on gelatinized 6-well plates, starting at $\approx 10^5$ cells. Medium was refreshed after 24 hours. Reporter plasmids were diluted 1:25 in pUC19b carrier plasmid (NEB) and mixed with Lipo-fectamine 2000 (Invitrogen). 16 µl reagent with 4 µg DNA in 300 µl Opti-MEM was added to 2 ml 2i medium per well. 4 hours post transfection, cells were detached using Accutase (EMD Millipore), split 1:2 and passaged onto gelatinized 60 mm plates in 3 ml 2i medium containing 3 μg doxycycline. Medium was refreshed 24 hours after passaging.

Flow cytometry

Cells were assayed on a LSRFortessa analyzer (BD Biosciences) two days after transfection. Cells were gated according to their forward (FSC-A) and side (SSC-A) scatter profiles. Each set of experiments contained at least one cell population transfected with the corresponding unregulated reporter construct and one mock transfected cell population (pUC19b carrier plasmid only), which was used to characterize background fluorescence.

Transcriptome profiling

Cells were transfected with reporter plasmid (as described above). Cells were sorted into four fractions (~100.000 cells each) on a FACSAriaIII cell sorter (BD Biosciences) according to ZsGreen intensities (fig. S10A). RNA from cells in each fraction was extracted using Trizol LS (Life Technologies). Sequencing libraries were prepared from isolated RNA using Illumnia TrueSeq Stranded mRNA kit. Libraries were sequences on an Illumnia HiSeq 2500 sequencer. Paired-end reads were mapped to RefSeq mRNA sequences (mm10) and mCherry mRNA sequence using Bowtie v2.2.0 (Langmead and Salzberg, 2012). Fragments per kilobase gene model per million mapped fragments (FPKM) was calculated for all transcripts and transcript isoforms were then aggregated to GeneSymbols. For further analysis we only considered genes expressed above 0.1 RPKM.

Taqman microRNA expression measurements

RNA was isolated from mESC V19 cells using Life Technologies miRVANA miRNA Isolation Kit. Expression of microRNAs mmu-miR-16, mmu-miR-20a and mmu-miR-290 was assayed using Life Technologies Taqman microRNA assays.

Flow cytometry data processing

For uni-regulated constructs (3'UTR only behind mCherry), cells were binned according to ZsGreen intensities (bin-width 0.2 in log10 space). The lower bin limit was set to the 0.9999-quantile of the background distribution. In each bin, outlier cells below 0.001-quantile and above 0.999-quantile of mCherry intensity distribution were discarded. 1,000 iterations of 50% subsampling were used to evaluate uncertainty of the data in each bin. From each subsampling iteration, mean and noise of mCherry intensities were calculated. Average of mean and noise values over all 1,000 iterations serve as observables for each particular bin. Standard deviation of mean and noise values over all 1,000 iterations serve as uncertainty of the observables for each particular bin.

For bi-regulated constructs (identical 3'UTRs behind ZsGreen and mCherry), cells were binned according to the summed [ZsGreen + mCherry] intensity (bin-width 0.2 in log10 space). The lower bin limit was set to the 0.9999-quantile of the summed [ZsGreen + mCherry] intensity of the background distribution. In each bin, ZsGreen intensity was normalized such that ZsGreen and mCherry intensity distributions had identical means. Mean and subsampled standard deviations for intrinsic noise were calculated in each bin by subsampling as described above. Intrinsic noise was calculated as $\eta_{int} = \sqrt{\frac{\overline{(z-m)^2}}{2 \cdot \bar{z} \cdot \bar{m}}}$ (Elowitz et al., 2002), with z and m as ZsGreen and mCherry intensities of cells and $\bar{x}$ denoting the mean of a variable over all cells in the bin. The aforementioned observables describe mean and noise of the measured fluorescences intensities. Biological signal mean and noise levels were deconvoluted from measured observables as described in Method Section 4.3.

Model fit to signal mean and noise

For all model fits to single cell data, a MATLAB implementation of the profile likelihood approach (Raue et al., 2009) was used to determine optimal fits and 95% confidence intervals of parameter estimates.

Uni-regulated reporter constructs (3'UTR only behind mCherry):

First, the bin-wise mean mCherry intensities of the regulated ($[M_{reg}]$) and corresponding unregulated ($[M_{unreg}]$) reporter constructs were fitted to the following equation

$$[M_{reg}] = \frac{1}{2} \cdot \left([M_{unreg}] - \mu^* \cdot K_M^* - K_M^* + \sqrt{([M_{unreg}] - \mu^* \cdot K_M^* - K_M^*)^2 + 4 \cdot [M_{unreg}] \cdot K_M^*} \right)$$

$$(26)$$

to obtain μ^*, a scaled total microRNA expression,

$$\mu^* = \frac{k^p}{d^p} \cdot \frac{d^m + d^{m\mu}}{d^m} \cdot [\mu^T] \quad , \tag{27}$$

and K_M^*, a scaled Michaelis-Menten constant,

$$K_M^* = \frac{k^p}{d^p} \cdot K_M \quad . \tag{28}$$

Next, repression R and saturation S were calculated as

$$R = 1 - \frac{1}{1 + \frac{\mu^*}{K_M^* + [M_{reg}]}} \quad , \tag{29}$$

and

$$S = \frac{[M_{reg}]}{K_M^* + [M_{reg}]} \quad . \tag{30}$$

Finally, the bin-wise total noise data from regulated and corresponding unregulated reporter constructs were simultaneously fit to the following noise equations

$$\eta_M^{unreg} = \sqrt{\frac{x}{[M_{unreg}]} + (\eta_{ext}^{unreg})^2}$$

$$\eta_M^{reg} = \sqrt{\frac{x}{[M_{reg}]} \cdot \frac{1-R}{(1-R \cdot S)^2} + \tilde{\eta}_\mu^2 \cdot \left(\frac{R}{1-R \cdot S}\right)^2 + (\eta_{ext}^{unreg})^2} \quad . \tag{31}$$

The fit yielded parameter estimates for the scaling factor $x = \frac{\overline{mCherry}}{[p]/b_0}$, which relates mean mCherry signal intensity $\overline{mCherry}$ to protein molecule numbers, the microRNA-independent extrinsic noise η_{ext}^{unreg} and the effective microRNA pool noise $\tilde{\eta}_\mu$. To estimate agreement of intrinsic noise reduction between our mechanistic model and the experimental noise data (as presented in Figure 15C) an additional fit parameter was multiplied to the first term in equation (31).

Bi-regulated constructs (identical 3'UTRs behind ZsGreen and mCherry): Intrinsic signal noise was fitted as proportional to the square root of summed [ZsGreen + mCherry] mean intensities $\overline{z + m}$ as $\eta_{int} = \frac{y}{\sqrt{\overline{z+m}}}$, with y as a scaling factor. Scaling factors for both the regulated reporters and the respective unregulated control reporters were estimated and their ratio yielded the intrinsic noise reduction conferred by microRNA regulation.

Mixed microRNA pool noise for correlated individual microRNA pools

The hypothetical microRNA pool noise of fully correlated individual microRNA pools (Figure 17C) was calculated as $\eta_{(x+y)} = \frac{\sqrt{\sigma_x^2 + \sigma_y^2 + 2 \cdot \sigma_x \cdot \sigma_y}}{s_x + s_y}$. The standard deviation σ was calculated as $\sigma_i = \eta_i \cdot s_i$, the product of noise in the individual microRNA pool η_i (known from mCherry reporters only regulated by the specific microRNA) times the relative microRNA pool size s_i, which we measured using Taqman microRNA assays.

Mapping flow cytometry experiments to transcriptome expression

To calculate the conversion factor from mCherry fluorescence intensities to FPKM values a least-square fit of respective values over the four bins was performed. Comparability of transcriptome expression from different bins is given by high similarity ($R^2 > 0.96$, fig. S10B). Relative effects of microRNA regulation on total noise as a function of FPKM values were calculated based on the parameters obtained from model fits to noise data from endogenous 3'UTRs (n = 3) and the mCherry fluorescent intensity to FPKM conversion factor.

Dicer knock-out mESC transcriptome expression data

Microarray expression data and Ago2 CLIP-seq data from wild-type and Dicer knock-out mouse embryonic stem cells (Leung et al., 2011) were obtained from Gene Expression Omnibus GSE25310 and data was processed as described in Leung et al. (2011). MicroRNA-mediated repression was calculated as the fold-change in mean expression between wild-type and Dicer knock-out samples. Loess regression was performed to obtain an error model relating standard deviations of expression for each gene as a function of mean expression over three replicates for both wild-type and knock-out samples. Significance of fold-changes was assessed at $\alpha < 0.05$ (Bonferroni corrected) by calculating z-scores as $z = \frac{\overline{x_{ko}} - \overline{x_{wt}}}{\sqrt{\sigma_{ko} \cdot \sigma_{wt}}}$, with $\overline{x_{ko}}$ and $\overline{x_{wt}}$ as mean signal intensities over three replicates in Dicer knockout and wild-type cells and σ as Loess estimated standard deviation at mean signal intensities over three replicates. Genes below log2 signal intensity of ~4.2 (40% of genes) were discarded as background. Genes were classified as 'supported' if at least one read cluster in their 3'UTR was found in CLIP-seq data and they contain at least one predicted conserved microRNA binding site (Targetscan6.2, (Garcia et al., 2011)) for a microRNA seed family expressed above 0.1% of total microRNA expression in mESC (Marson et al., 2008).

4.3 Measuring fluorescent protein variability on a flow cytometry

Quantitative flow cytometry requires compensation for both background fluorescence and instrument noise. While these effects are easily mitigated for mean signal estimates, the relationship between the variance measured on a flow cytometer and the true signal variance is nontrivial. The principal difficulty is that cell size variation and instrument noise induce a scale dependent correlation of the signal and background, which increases the measured variance of their sum (intuitively, consider the increased likelihood of extreme values being simultaneously realized). As a consequence, the total measured variance provides an inflated estimate of the true signal variance. Here we address this challenge by providing an exact analytic framework for deconvoluting variation in fluorescent protein abundance from these artifactual noise sources.

We begin our analysis by calculating the effect of fluorescent cellular background on both the measured channel noise as well as the covariance between channels. For random variables x and x', let $\eta_x^2 = \text{Var}(x)/\langle x \rangle^2$ and $\eta_{x,x'}^2 = \text{Cov}(x,x')/\langle x \rangle \langle x' \rangle$ denote the normalized variance and normalized covariance, respectively. Consider the following affine stochastic model for a flow cytometry measurement t of a fluorescent signal s:

$$t = s + \theta, \tag{32}$$

where θ is a random variable representing the fluorescent background of a single cell. Since many cellular signals – including autofluorescence as well as RNA and protein abundance – scale with cell size, the background is often correlated with the signal of interest. This size-induced correlation between the background and signal distorts the measurement of the signal variance as shown below:

$$\text{Var}(s) = \text{Var}(t) - \text{Var}(\theta) - 2 \cdot \text{Cov}(s,\theta). \tag{33}$$

While the background variance can be measured directly, the size-induced covariance distortion is dependent on the particular signal being measured. In this work, we consider cellular signals which correspond to molecular species whose average concentration is invariant of cell size. This is a broad class of signals, capturing most cell-cycle invariant transcriptome and proteome measurements. Formally, a signal s is admissible within this class if it satisfies the following scaling property with respect to the cell size ω:

$$\frac{\partial}{\partial \boldsymbol{\omega}} \mathsf{E}\left[s/\boldsymbol{\omega}|\boldsymbol{\omega}=\omega\right] = 0. \tag{34}$$

If the signal and background are correlated exclusively through cell size, then for each signal in this class, the following relation holds:

$$\mathsf{Cov}(s, \boldsymbol{\theta}) = \langle s \rangle \langle \boldsymbol{\theta} \rangle \eta_\omega^2. \tag{35}$$

More concisely, the normalized covariance of the signal and background is equal to the normalized variance in cell size: $\eta_{s,\theta}^2 = \eta_\omega^2$. This covariance scaling can be directly validated by measuring the covariance between a flow cytometry channel (excitation laser and detection filter pair) which measures an expressed fluorescent protein ($t = s + \boldsymbol{\theta}$) and a spectrally isolated channel which does not detect the protein (ϕ). The model predicts that a spectrally isolated channel will be correlated with the signal through cell size:

$$\mathsf{Cov}(t, \phi) = \mathsf{Cov}(\boldsymbol{\theta}, \phi) + \langle s \rangle \langle \phi \rangle \eta_\omega^2. \tag{36}$$

Both the predicted covariance offset and mean scaling is precisely recapitulated by GFP protein expression measurements across a broad dynamic range (Figure 33, data adapted from Klemm et al. (2014)). We further performed the same analysis for GFP mRNA using a recently developed method for measuring fluorescently labeled RNA on a flow cytometer (Klemm et al., 2014). These data not only recapitulate the predicted scaling phenomenon, but also quantitatively yield an identical estimate of the cell size noise η_ω^2 (Figure 33).

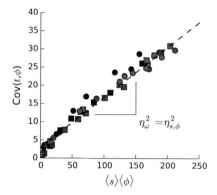

Figure 33: **Cell size mediated background and signal covariance.**
Ectopic GFP expressed at doxycycline controlled induction levels in mouse embryonic stem cells. Protein (measured with excitation laser Ex. 488nm, and emission Em. 505-525nm) and RNA (Ex. 635nm, Em. 655-685nm) covariance measurements are reported as squares and circles, respectively. (black) Covariance scaling of protein and RNA signals with a spectrally isolated background channel (Ex. 405nm, Em. 785-815nm). (blue) Biological replicate. (red) Technical replicate covariance measurements for an alternate, spectrally isolated channel (SSC). Collectively these data demonstrate linearly covariance scaling with respect to a signal independent cell-size noise parameter $\eta_\omega^2 \approx 0.07$.

These data establish the model assumptions implicit in equations (32) and (34). Consequently the normalized signal variance for signals satisfying (34) is derived from (33) and (35) as

$$\eta_s^2 = \left(\frac{\langle t \rangle}{\langle s \rangle}\right)^2 \eta_t^2 - \left(\frac{\langle \theta \rangle}{\langle s \rangle}\right)^2 \eta_\theta^2 - 2 \cdot \left(\frac{\langle \theta \rangle}{\langle s \rangle}\right) \eta_\omega^2. \tag{37}$$

This model based calculation of the normalized signal variance η_s^2 reflects a correction to the total normalized variance η_t^2. Importantly, this analysis shows the profound effect of the background on the total noise measurement, including $\langle s \rangle^{-2}$ and $\langle s \rangle^{-1}$ distortion modes.

For measurements satisfying (32) and (34), the intrinsic noise $I_{s|s'} = \eta_s^2 - \eta_{s,s'}^2$ of signal s with respect to signal s' is given by

$$I_{s|s'} = \frac{1}{\langle s \rangle^2}\left[V_{t|t'} - V_{\theta|\theta'}\right] - \frac{\Delta(\theta, \theta')}{\langle s \rangle}\eta_\omega^2. \tag{38}$$

where $V_{X|X'}$ denotes the intrinsic variance $\mathrm{Var}(X) - \mathrm{Cov}(X, X')$ for any random variable X, $\Delta(\theta, \theta') = \langle \theta \rangle - \langle \theta' \rangle$, and without loss of generality $\langle s \rangle = \langle s' \rangle$. For the case of the symmetric, bidirectional expression system used in this work, a simplified intrinsic noise estimate is given as

$$I_{s|s'} = \left(\frac{\langle t \rangle}{\langle s \rangle} \right)^2 I_{t|t'} - \left(\frac{\langle \theta \rangle}{\langle s \rangle} \right)^2 I_{\theta|\theta'}, \tag{39}$$

where $I_{t|t'}$ and $I_{\theta|\theta'}$ are the total and background intrinsic noise measurements, respectively.

An additional source of measurement variability arises from detection noise in the flow cytometer, which is captured by the following extended model:

$$t = \xi \cdot s + \theta, \tag{40}$$

where ξ is the channel dependent detection efficiency of the flow cytometer and θ is the measured background fluorescence. (In this notation, $\theta = \xi \cdot \theta'$, where θ' is the true background fluorescence.) The detection efficiency ξ is taken to have unit mean and is independent between different laser channels since these measurements use different detectors and are acquired sequentially at different physical positions in the flow stream. For the more general model in (40) the intrinsic noise is given by

$$I_{s|s'} = \frac{\langle t \rangle^2}{\langle s \rangle^2} \left(\frac{\eta_t^2}{1 + \eta_\xi^2} - \eta_{t,t'}^2 \right) - \frac{\langle \theta \rangle^2}{\langle s \rangle^2} \left(\frac{\eta_\theta^2}{1 + \eta_\xi^2} - \eta_{\theta,\theta'}^2 \right) - \left(1 + 2\frac{\langle \theta \rangle}{\langle s \rangle} \right) \frac{\eta_\xi^2}{1 + \eta_\xi^2}. \tag{41}$$

For the flow cytometry measurements in this work, $\eta_\xi^2 \ll 1$; in this case, the intrinsic signal noise is estimated in terms of the intrinsic channel and background noise components as

$$I_{s|s'} \approx \left(\frac{\langle t \rangle}{\langle s \rangle} \right)^2 I_{t|t'} - \left(\frac{\langle \theta \rangle}{\langle s \rangle} \right)^2 I_{\theta|\theta'} - \left(1 + 2\frac{\langle \theta \rangle}{\langle s \rangle} \right) \eta_\xi^2. \tag{42}$$

While the measured intrinsic channel noise is independent of extrinsic cell size heterogeneity, explicit compensation is required for both the background and detection efficiency noise contributions. Importantly the technical detection noise contributes to persistent intrinsic noise between channels even in the limit of high mean expression levels. This intrinsic noise offset is clearly observed even in the absence of intrinsic signal noise ($I_{s|s'} = 0$), when an identical signal is measured by two independent channels with different lasers and detection filters (Figure 34).

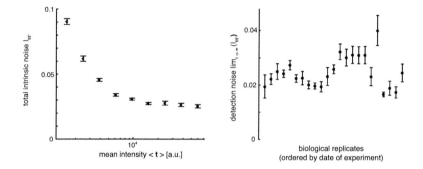

Figure 34: **Persistent total intrinsic noise between channels measuring identical fluorescent reporter**
Total intrinsic noise scaling with mean intensity between two independent channel measurements (Ex. 488nm, Em. 695 LP and Ex. 561nm, Em. 610/20) of mCherry fluorescent protein (**left**). Reproducible estimates of the detection noise for a broad range of samples (**right**). Dots and error bars represent mean and bootstrapped standard deviation of estimates (**left**). Dots and error bars represent optimal model fit and 95% confidence interval of model fit (**right**).

In this case, the persistent total intrinsic noise $I_{t|t'}$ matches the prediction (see equation (42)) that $\lim_{\langle t \rangle \to \infty} I_{t|t'} \approx \eta_\xi^2$. The total signal noise under the generalized model in (40) is then given as

$$
\eta_s^2 = \left(\frac{1}{1+\eta_\xi^2}\right)\left(\frac{\langle t \rangle}{\langle s \rangle}\right)^2 \eta_t^2 - \left(\frac{1}{1+\eta_\xi^2}\right)\left(\frac{\langle \theta \rangle}{\langle s \rangle}\right)^2 \eta_\theta^2 - 2 \cdot \frac{\langle \theta \rangle}{\langle s \rangle}\eta_\omega^2 - \left(1 + 2\frac{\langle \theta \rangle}{\langle s \rangle}\right)\frac{\eta_\xi^2}{1+\eta_\xi^2} \tag{43}
$$

Together, these calculations establish an analytic framework for quantifying variability in fluorescent protein abundance using a flow cytometer. The proposed corrective noise model depends only on parameters which are directly estimated from raw flow cytometry data. Consequently, the approach yields unbiased noise estimates without requiring additional calibration experiments or interfering with downstream statistical analysis.

The necessity for deconvolution of biological signal variation from technical flow cytometry noise is apparent when calculating the contributions of different artifactual noise sources to total noise η_t^2 (Figure 35).

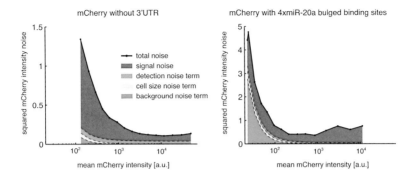

Figure 35: **Contributions of signal noise, detection noise, cell size noise and background noise to total noise η_t^2.**
(**left**) Unregulated mCherry reporter without 3'UTR with mean intensities far above background fluorescence. (**right**) Regulated mCherry reporter with four bulged miR-20a binding sites with low mean intensities close to background fluorescence.

This is further exemplified by a sensitivity analysis of the agreement between the theoretical predictions and experimental estimates of the extent of microRNA-mediated intrinsic noise reduction (Figure 36). Because intrinsic noise is dominant at low expression levels these estimates are very sensitive to proper deconvolution of biological signal variation from technical flow cytometry noise.

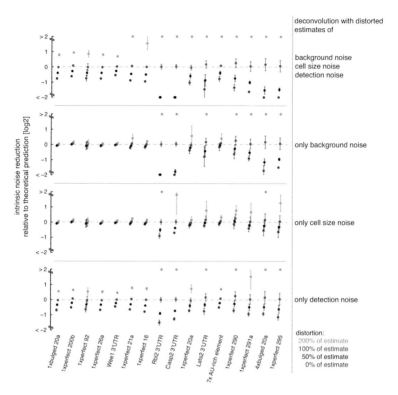

Figure 36: **Sensitivity analysis for intrinsic noise reduction estimates.**
Plots show agreement of experimentally derived intrinsic noise reduction estimates with theoretical prediction for all measured reporter constructs. Agreement is accurate when all noise sources are correctly deconvoluted according to the model (red). Over- (green) or under-estimates (black and blue) of artifactual noise sources leads to over- or underestimates of the experimentally observed amount of intrinsic noise reduction compared to theoretical prediction, respectively. For all measurements $n \geq 3$, except 7xAU-rich element $n = 2$.

4.4 GO Term microRNA regulation

'Cellular Component' GO terms *enriched* for human microRNA binding sites

Summary	Pos.	GO Term	GO name	log10 (q-value)	median(cons. miRNA sites)	#genes
membrane/	1	GO:0030054	cell junction	-12.4	20.5	388
cell-cell	12	GO:0005911	cell-cell junction	-4.8	23	107
	30	GO:0014704	intercalated disc	-3.5	25.5	34
	33	GO:0005913	cell-cell adherens junction	-3.2	31	32
	37	GO:0005925	focal adhesion	-2.4	17	131
neuronal	2	GO:0045202	synapse	-11.4	34	150
	3	GO:0030425	dendrite	-8	21	233
	4	GO:0014069	postsynaptic density	-6.3	23.5	108
	5	GO:0030424	axon	-5.8	19	161
	7	GO:0043025	neuronal cell body	-5.4	19	255
	8	GO:0030027	lamellipodium	-5	20	125
	10	GO:0045211	postsynaptic membrane	-4.9	18.5	190
	13	GO:0043005	neuron projection	-4.8	21.5	160
	15	GO:0030175	filopodium	-4.6	26	55
	19	GO:0042734	presynaptic membrane	-4.4	26	50
	20	GO:0043198	dendritic shaft	-4.4	31	33
	22	GO:0030426	growth cone	-4.1	23	95
	23	GO:0043197	dendritic spine	-3.8	27	78
	24	GO:0030672	synaptic vesicle membrane	-3.8	31	45
	38	GO:0060076	excitatory synapse	-2.4	44	17
	44	GO:0030673	axolemma	-2.2	44	14
	48	GO:0031594	neuromuscular junction	-2.1	21.5	46
	50	GO:0001726	ruffle	-2	18.5	86
organelle	6	GO:0005794	Golgi apparatus	-5.5	14	735
	14	GO:0005769	early endosome	-4.6	21	148
	16	GO:0000139	Golgi membrane	-4.5	14	473
	26	GO:0010494	cytoplasmic stress granule	-3.7	36	30
	27	GO:0005802	trans-Golgi network	-3.7	25.5	106
	35	GO:0030136	clathrin-coated vesicle	-2.8	33	41
	46	GO:0005768	endosome	-2.1	14	163

macromolecular complexes	9	GO:0005667	transcription factor complex	-5	20	218
	21	GO:0017053	transcriptional repressor complex	-4.1	40	52
	25	GO:0032993	protein-DNA complex	-3.8	54	14
	31	GO:0043234	protein complex	-3.5	16.5	260
	34	GO:0000151	ubiquitin ligase complex	-2.8	25.5	62
	36	GO:0005875	microtubule associated complex	-2.7	38.5	30
	40	GO:0030877	beta-catenin destruction complex	-2.3	51	10
	42	GO:0016580	Sin3 complex	-2.2	44.5	12
	43	GO:0005891	voltage-gated calcium channel complex	-2.2	28	28
	45	GO:0000118	histone deacetylase complex	-2.2	42	29
	47	GO:0035098	ESC/E(Z) complex	-2.1	53	13
	49	GO:0016442	RISC complex	-2.1	48.5	10
nucleus	11	GO:0000790	nuclear chromatin	-4.8	22	94
	17	GO:0000785	chromatin	-4.5	21	73
	28	GO:0031519	PcG protein complex	-3.7	42	24
	29	GO:0005654	nucleoplasm	-3.7	11	1071
	39	GO:0000792	heterochromatin	-2.3	38.5	18
	51	GO:0005730	nucleolus	-2	10	1707
cytoplasm	18	GO:0030659	cytoplasmic vesicle membrane	-4.4	20	108
	32	GO:0048471	perinuclear region of cytoplasm	-3.3	14	505
	41	GO:0015629	actin cytoskeleton	-2.2	16	190

'Molecular Function' GO terms *enriched* for human microRNA binding sites

Summary	Pos.	GO Term	GO name	log10 (q-value)	median(cons. miRNA sites)	#genes
transcriptional	1	GO:0003682	chromatin binding	-11.9	21	331
regulation	2	GO:0008134	transcription factor binding	-9.9	23	268
	3	GO:0043565	sequence-specific DNA binding	-9.1	19	521
	6	GO:0044212	transcription regulatory region DNA binding	-6.2	24	167
	7	GO:0003700	sequence-specific DNA binding transcription factor activity	-5	14	999
	11	GO:0001077	RNA polymerase II core promoter proximal region sequence-specific DNA binding transcription factor activity involved in positive regulation of transcription	-4.7	35	76
	13	GO:0003714	transcription corepressor activity	-4.5	22	181
	14	GO:0001105	RNA polymerase II transcription coactivator activity	-3.6	71.5	22
	16	GO:0001078	RNA polymerase II core promoter proximal region sequence-specific DNA binding transcription factor activity involved in negative regulation of transcription	-3.1	31.5	34
	20	GO:0000979	RNA polymerase II core promoter sequence-specific DNA binding	-2.4	32	29

	22	GO:0000987	core promoter proximal region sequence-specific DNA binding	-2.2	65.5	16
	24	GO:0042826	histone deacetylase binding	-2.2	32	67
	27	GO:0000978	RNA polymerase II core promoter proximal region sequence-specific DNA binding	-2.1	29	42
	28	GO:0003713	transcription coactivator activity	-2.1	18	232
post-transcriptional regulation	21	GO:0003730	mRNA 3'-UTR binding	-2.2	33	27
protein binding/ signaling	4	GO:0005515	protein binding	-9	12	5918
	5	GO:0004842	ubiquitin-protein ligase activity	-6.2	23	256
	8	GO:0008013	beta-catenin binding	-4.9	39.5	62
	9	GO:0019901	protein kinase binding	-4.9	17	299
	10	GO:0004674	protein serine/threonine kinase activity	-4.8	15.5	346
	12	GO:0019904	protein domain specific binding	-4.6	22	173
	15	GO:0044325	ion channel binding	-3.5	26.5	78
	17	GO:0031625	ubiquitin protein ligase binding	-3.1	19	159
	19	GO:0046332	SMAD binding	-2.7	44	40
	23	GO:0030165	PDZ domain binding	-2.2	18.5	92
	25	GO:0046875	ephrin receptor binding	-2.2	39	29
	26	GO:0048185	activin binding	-2.2	61.5	12
	18	GO:0005024	transforming growth factor beta-activated receptor activity	-2.7	75	15

'Biological Pathway' GO terms *enriched* for human microRNA binding sites

Summary	Pos.	GO Term	GO name	log10 (q-value)	median(cons. miRNA sites)	#genes
regulation of transcription	2	GO:0045944	positive regulation of transcription from RNA polymerase II promoter	-11.2	20	729
	3	GO:0045893	positive regulation of transcription, DNA-templated	-10.8	21.5	498
	4	GO:0000122	negative regulation of transcription from RNA polymerase II promoter	-9.2	21	555
	8	GO:0045892	negative regulation of transcription, DNA-templated	-5.5	21	439
	19	GO:0000381	regulation of alternative mRNA splicing, via spliceosome	-3	56.5	26
	30	GO:0016568	chromatin modification	-2.4	26	99
nervous system development	1	GO:0007411	axon guidance	-12.2	25	333
	12	GO:0007399	nervous system development	-4.6	24	275
	17	GO:0001764	neuron migration	-3.7	26.5	102
	18	GO:0048169	regulation of long-term neuronal synaptic plasticity	-3.4	61.5	20
	20	GO:0001843	neural tube closure	-3	26.5	68
	23	GO:0045665	negative regulation of neuron differentiation	-2.7	27.5	46
	32	GO:0007420	brain development	-2.4	18	175
	33	GO:0048666	neuron development	-2.4	34	36
	34	GO:0048699	generation of neurons	-2.3	74	10
	35	GO:0014047	glutamate secretion	-2.2	38.5	18
	36	GO:0030182	neuron differentiation	-2.2	25	82
	37	GO:0048813	dendrite morphogenesis	-2.2	39	32
	43	GO:0030900	forebrain development	-2.1	29	55

89

	46	GO:0010976	positive regulation of neuron projection development	-2	26	61
development	9	GO:0001701	in utero embryonic development	-5.2	23	218
	10	GO:0060021	palate development	-5	32.5	68
	11	GO:0009952	anterior/posterior pattern specification	-4.6	30	87
	24	GO:0009791	post-embryonic development	-2.7	26	79
	25	GO:0002053	positive regulation of mesenchymal cell proliferation	-2.5	44	35
	26	GO:0007507	heart development	-2.5	22	149
	29	GO:0009954	proximal/distal pattern formation	-2.5	37	23
	39	GO:0030879	mammary gland development	-2.2	31.5	22
	40	GO:0001702	gastrulation with mouth forming second	-2.1	55.5	22
	41	GO:0016477	cell migration	-2.1	20	125
	45	GO:0001525	angiogenesis	-2	18	205
signaling	5	GO:0048011	neurotrophin TRK receptor signaling pathway	-6.9	22	272
	6	GO:0007173	epidermal growth factor receptor signaling pathway	-6	23.5	190
	7	GO:0008543	fibroblast growth factor receptor signaling pathway	-5.6	28.5	160
	13	GO:0038095	Fc-epsilon receptor signaling pathway	-4.6	23	165
	14	GO:0016055	Wnt signaling pathway	-4.1	21	155
	15	GO:0090090	negative regulation of canonical Wnt signaling pathway	-4	31	83
	21	GO:0007179	transforming growth factor beta receptor signaling pathway	-3	21	123
	22	GO:0060070	canonical Wnt signaling pathway	-3	27.5	76

	28	GO:0008286	insulin receptor signaling pathway	-2.5	22	148
	31	GO:0048013	ephrin receptor signaling pathway	-2.4	40	29
	42	GO:0048008	platelet-derived growth factor receptor signaling pathway	-2.1	33	27
	44	GO:0007219	Notch signaling pathway	-2	21	121
protein modification	16	GO:0006468	protein phosphorylation	-3.8	18	340
	27	GO:0018107	peptidyl-threonine phosphorylation	-2.5	47	31
	38	GO:0006470	protein dephosphorylation	-2.2	23.5	106

'Cellular Component' GO terms *depleted* for human microRNA binding sites

Summary	Pos.	GO Term	GO name	log10 (q-value)	median(cons. miRNA sites)	#genes
extracellular / secretion	1	GO:0005576	extracellular region	-54.4	3	1430
	3	GO:0005615	extracellular space	-20.7	4	1014
	11	GO:0072562	blood microparticle	-8	2	106
	22	GO:0034364	high-density lipoprotein particle	-2.6	1	20
	25	GO:0030141	secretory granule	-2.4	2.5	72
	27	GO:0070062	extracellular vesicular exosome	-2.4	7	1545
membrane	2	GO:0016021	integral component of membrane	-25.9	6	4006
	12	GO:0005886	plasma membrane	-7.8	8	3454
	23	GO:0001533	cornified envelope	-2.5	2	21
mitochondria	4	GO:0005739	mitochondrion	-18.8	5	1186
	9	GO:0005743	mitochondrial inner membrane	-10	4	324
	15	GO:0005747	mitochondrial respiratory chain complex I	-6.4	1	39
	16	GO:0005759	mitochondrial matrix	-6.2	4	234
	19	GO:0005761	mitochondrial ribosome	-4.3	0.5	24
	24	GO:0005762	mitochondrial large ribosomal subunit	-2.4	0	15
filament	5	GO:0045095	keratin filament	-18.6	1	94
	8	GO:0005882	intermediate filament	-11	1	106
ribosome	6	GO:0005840	ribosome	-17.1	1	139
	7	GO:0022625	cytosolic large ribosomal subunit	-11.5	0	52
	10	GO:0022627	cytosolic small ribosomal subunit	-9	0	35
	14	GO:0015935	small ribosomal subunit	-7.1	0	25
misc	18	GO:0031514	motile cilium	-4.5	1	41
	20	GO:0005929	cilium	-2.9	3	102
	21	GO:0042613	MHC class II protein complex	-2.7	1	16
	26	GO:0036128	CatSper complex	-2.4	0	9

28	GO:0000786	nucleosome	-2	3	56
29	GO:0005833	hemoglobin complex	-2	1	11

'Molecular Function' GO terms *depleted* for human microRNA binding sites

Summary	Pos.	GO Term	GO name	log10 (q-value)	median(cons. miRNA sites)	#genes
sensory perception	1	GO:0004984	olfactory receptor activity	-151.9	0	365
	2	GO:0004930	G-protein coupled receptor activity	-126.4	0	592
	15	GO:0008527	taste receptor activity	-4.3	0	15
	25	GO:0004982	N-formyl peptide receptor activity	-2.1	0	9
ribosome	3	GO:0003735	structural constituent of ribosome	-31.1	0	150
protein degradation	4	GO:0004252	serine-type endopeptidase activity	-11.1	2	152
	5	GO:0004867	serine-type endopeptidase inhibitor activity	-8.7	1	92
	20	GO:0004869	cysteine-type endopeptidase inhibitor activity	-3.6	0.5	30
	26	GO:0004866	endopeptidase inhibitor activity	-2.1	1	33
metabolism	9	GO:0020037	heme binding	-5.7	2.5	128
	10	GO:0030246	carbohydrate binding	-5.2	3	143
	11	GO:0008137	NADH dehydrogenase (ubiquinone) activity	-5.2	1	38
	12	GO:0005506	iron ion binding	-5	3	148
	13	GO:0016491	oxidoreductase activity	-5	4	172
	14	GO:0019825	oxygen binding	-4.4	1	33
	16	GO:0003824	catalytic activity	-4.2	4	211
	17	GO:0004497	monooxygenase activity	-4	1	39
	18	GO:0009055	electron carrier activity	-3.8	2	74
	21	GO:0004364	glutathione transferase activity	-3.5	1	24
	22	GO:0070330	aromatase activity	-3.3	0.5	22
	23	GO:0015020	glucuronosyltransferase activity	-2.6	3	24

	24	GO:0016705	oxidoreductase activity, acting on paired donors, with incorporation or reduction of molecular oxygen	-2.6	2	37
immune response	19	GO:0005132	type I interferon receptor binding	-3.6	0	17
	27	GO:0005125	cytokine activity	-2	5	164
	29	GO:0046703	natural killer cell lectin-like receptor binding	-2	0	7
misc	7	GO:0005179	hormone activity	-7.5	2	93
	8	GO:0003676	nucleic acid binding	-5.7	5	768
	28	GO:0019843	rRNA binding	-2	1	27

'Biological Function' GO terms *depleted* for human microRNA binding sites

Summary	Pos.	GO Term	GO name	log10 (q-value)	median(cons. miRNA sites)	#genes
translation	1	GO:0006412	translation	-26.6	1	246
	4	GO:0006414	translational elongation	-23.3	0	88
	5	GO:0006415	translational termination	-22.1	0	83
	6	GO:0006614	SRP-dependent cotranslational protein targeting to membrane	-19	0	105
	7	GO:0006413	translational initiation	-17.2	0	116
	14	GO:0044267	cellular protein metabolic process	-11.3	5	519
	23	GO:0006364	rRNA processing	-8.8	2	93
	55	GO:0006418	tRNA aminoacylation for protein translation	-4.1	3	41
	77	GO:0008033	tRNA processing	-3.3	2	48
	94	GO:0042274	ribosomal small subunit biogenesis	-2.5	0.5	12
	100	GO:0051084	'de novo' posttranslational protein folding	-2.3	3	38
	108	GO:0051258	protein polymerization	-2.1	2	28
immune system	2	GO:0019083	viral transcription	-24.4	0	80
	8	GO:0042742	defense response to bacterium	-16.2	1	101
	10	GO:0006955	immune response	-15.1	4	322
	11	GO:0019058	viral life cycle	-14.4	1	114
	22	GO:0051607	defense response to virus	-9	3	143
	28	GO:0006954	inflammatory response	-7.8	5	301
	32	GO:0050776	regulation of immune response	-6.7	2	76
	34	GO:0042590	antigen processing and presentation of exogenous peptide antigen via MHC class I	-5.9	3	78

35	GO:0002479	antigen processing and presentation of exogenous peptide antigen via MHC class I, TAP-dependent	-5.9	2.5	74
36	GO:0006952	defense response	-5.9	2	64
37	GO:0060337	type I interferon signaling pathway	-5.5	1	65
42	GO:0016032	viral process	-5	6	534
46	GO:0006968	cellular defense response	-4.6	2	58
47	GO:0045071	negative regulation of viral genome replication	-4.6	0	34
48	GO:0006935	chemotaxis	-4.6	5	115
51	GO:0050832	defense response to fungus	-4.2	0	18
54	GO:0019221	cytokine-mediated signaling pathway	-4.1	4	227
57	GO:0043330	response to exogenous dsRNA	-4	1	34
58	GO:0009615	response to virus	-4	4	113
60	GO:0002474	antigen processing and presentation of peptide antigen via MHC class I	-3.9	3	96
62	GO:0031640	killing of cells of other organism	-3.8	0	11
63	GO:0045343	regulation of MHC class I biosynthetic process	-3.7	0	17
65	GO:0002323	natural killer cell activation involved in immune response	-3.7	0	17
68	GO:0006956	complement activation	-3.6	1	25
72	GO:0002376	immune system process	-3.5	2	38
81	GO:0045087	innate immune response	-3	7	609
86	GO:0006953	acute-phase response	-2.9	1	37
87	GO:0006958	complement activation, classical pathway	-2.9	1.5	28

	88	GO:0002286	T cell activation involved in immune response	-2.8	1	21
	93	GO:0050830	defense response to Gram-positive bacterium	-2.5	3	36
	95	GO:0007597	blood coagulation, intrinsic pathway	-2.4	1	17
	109	GO:0030449	regulation of complement activation	-2	1.5	22
metabolism	3	GO:0044281	small molecule metabolic process	-23.6	6	1391
	9	GO:0006805	xenobiotic metabolic process	-15.5	1	143
	16	GO:0022904	respiratory electron transport chain	-11.1	1	86
	18	GO:0016070	RNA metabolic process	-10.3	2	242
	19	GO:0044237	cellular metabolic process	-9.5	2	128
	20	GO:0007586	digestion	-9.2	1	54
	25	GO:0034641	cellular nitrogen compound metabolic process	-8.8	3	185
	29	GO:0055114	oxidation-reduction process	-7.8	3	144
	31	GO:0008202	steroid metabolic process	-7.2	1	56
	39	GO:0019369	arachidonic acid metabolic process	-5.2	1	53
	43	GO:0006120	mitochondrial electron transport, NADH to ubiquinone	-4.8	1	39
	67	GO:0017144	drug metabolic process	-3.7	1	27
	70	GO:0006749	glutathione metabolic process	-3.5	1	34
	73	GO:0055086	nucleobase-containing small molecule metabolic process	-3.5	4	78

	74	GO:0006521	regulation of cellular amino acid metabolic process	-3.4	3	51
	80	GO:0016042	lipid catabolic process	-3.1	4	77
	83	GO:1901687	glutathione derivative biosynthetic process	-2.9	1	25
	90	GO:0006629	lipid metabolic process	-2.5	5	149
	91	GO:0006691	leukotriene metabolic process	-2.5	1	17
	96	GO:0019373	epoxygenase P450 pathway	-2.3	0	11
	97	GO:0006200	ATP catabolic process	-2.3	6	143
	101	GO:0008206	bile acid metabolic process	-2.3	3	36
	105	GO:0006081	cellular aldehyde metabolic process	-2.1	0	11
sensory perception	13	GO:0050911	detection of chemical stimulus involved in sensory perception of smell	-12.1	0	31
	27	GO:0007608	sensory perception of smell	-8.3	0	56
	50	GO:0050909	sensory perception of taste	-4.3	0	26
	69	GO:0050912	detection of chemical stimulus involved in sensory perception of taste	-3.6	0	10
	76	GO:0007601	visual perception	-3.3	5	188
	82	GO:0001580	detection of chemical stimulus involved in sensory perception of bitter taste	-3	0	16
signaling	15	GO:0007186	G-protein coupled receptor signaling pathway	-11.1	4	307
	56	GO:0007166	cell surface receptor signaling pathway	-4	5	159
	99	GO:0007218	neuropeptide signaling pathway	-2.3	4	98
	103	GO:0048387	negative regulation of retinoic acid receptor signaling pathway	-2.1	0	24

	110	GO:0007200	phospholipase C-activating G-protein coupled receptor signaling pathway	-2	2	41
protein degradation	24	GO:0010951	negative regulation of endopeptidase activity	-8.8	1	86
	33	GO:0030162	regulation of proteolysis	-6.2	1	38
	41	GO:0051436	negative regulation of ubiquitin-protein ligase activity involved in mitotic cell cycle	-5	2	64
	45	GO:0031145	anaphase-promoting complex-dependent proteasomal ubiquitin-dependent protein catabolic process	-4.6	3	79
	49	GO:0006508	proteolysis	-4.4	4.5	204
	52	GO:0051437	positive regulation of ubiquitin-protein ligase activity involved in mitotic cell cycle	-4.1	3	70
	59	GO:0051439	regulation of ubiquitin-protein ligase activity involved in mitotic cell cycle	-3.9	3	74
	85	GO:0033141	positive regulation of peptidyl-serine phosphorylation of STAT protein	-2.9	0.5	18
	102	GO:0051603	proteolysis involved in cellular protein catabolic process	-2.3	2.5	28
reproduction	30	GO:0007338	single fertilization	-7.7	1	50
	38	GO:0032504	multicellular organism reproduction	-5.2	0	23
	40	GO:0007283	spermatogenesis	-5.1	5	331
	61	GO:0007339	binding of sperm to zona pellucida	-3.8	1	34
	92	GO:0035036	sperm-egg recognition	-2.5	0	11
DNA repair	44	GO:0006281	DNA repair	-4.7	6	272

	66	GO:0006289	nucleotide-excision repair	-3.7	4	68
	78	GO:0006977	DNA damage response, signal transduction by p53 class mediator resulting in cell cycle arrest	-3.3	3	66
	89	GO:0006283	transcription-coupled nucleotide-excision repair	-2.6	3	46
	98	GO:0010529	negative regulation of transposition	-2.3	0	7
misc	17	GO:0000184	nuclear-transcribed mRNA catabolic process, nonsense-mediated decay	-10.5	1	116
	21	GO:0016071	mRNA metabolic process	-9.1	2	220
	26	GO:0031424	keratinization	-8.4	1	45
	53	GO:0001539	ciliary or bacterial-type flagellar motility	-4.1	0.5	18
	64	GO:0010467	gene expression	-3.7	7	666
	71	GO:0000278	mitotic cell cycle	-3.5	6	399
	75	GO:0071276	cellular response to cadmium ion	-3.3	0	14
	79	GO:0001895	retina homeostasis	-3.2	2	31
	84	GO:0006355	regulation of transcription, DNA-templated	-2.9	8	1219
	104	GO:0030049	muscle filament sliding	-2.1	2	38
	106	GO:0007204	positive regulation of cytosolic calcium ion concentration	-2.1	5	109
	107	GO:0071294	cellular response to zinc ion	-2.1	0	12

References

E. Alvarez-Saavedra and H. R. Horvitz. Many families of C. elegans microRNAs are not essential for development or viability. *Current biology : CB*, 20(4):367–373, Feb. 2010.

M. Ashburner, C. A. Ball, J. A. Blake, D. Botstein, H. Butler, J. M. Cherry, A. P. Davis, K. Dolinski, S. S. Dwight, J. T. Eppig, M. A. Harris, D. P. Hill, L. Issel-Tarver, A. Kasarskis, S. Lewis, J. C. Matese, J. E. Richardson, M. Ringwald, G. M. Rubin, and G. Sherlock. Gene ontology: tool for the unification of biology. The Gene Ontology Consortium. *Nature Genetics*, 25(1):25–29, May 2000.

A. Baccarini, H. Chauhan, T. J. Gardner, A. D. Jayaprakash, R. Sachidanandam, and B. D. Brown. Kinetic analysis reveals the fate of a microRNA following target regulation in mammalian cells. *Current biology : CB*, 21(5):369–376, Mar. 2011.

D. Baek, J. Villén, C. Shin, F. D. Camargo, S. P. Gygi, and D. P. Bartel. The impact of microRNAs on protein output. *Nature*, 455(7209):64–71, Sept. 2008.

A. G. Baltz, M. Munschauer, B. Schwanhäusser, A. Vasile, Y. Murakawa, M. Schueler, N. Youngs, D. Penfold-Brown, K. Drew, M. Milek, E. Wyler, R. Bonneau, M. Selbach, C. Dieterich, and M. Landthaler. The mRNA-bound proteome and its global occupancy profile on protein-coding transcripts. *Molecular Cell*, 46 (5):674–690, June 2012.

S. Bamford, E. Dawson, S. Forbes, J. Clements, R. Pettett, A. Dogan, A. Flanagan, J. Teague, P. A. Futreal, M. R. Stratton, and R. Wooster. The COSMIC (Catalogue of Somatic Mutations in Cancer) database and website. *British journal of cancer*, 91(2):355–358, July 2004.

A. Bar-Even, J. Paulsson, N. Maheshri, M. Carmi, E. O'Shea, Y. Pilpel, and N. Barkai. Noise in protein expression scales with natural protein abundance. *Nature Genetics*, 38(6):636–643, June 2006.

C. Barreau, L. Paillard, and H. B. Osborne. AU-rich elements and associated factors: are there unifying principles? *Nucleic Acids Research*, 33(22):7138–7150, 2005.

D. P. Bartel. MicroRNAs: genomics, biogenesis, mechanism, and function. *Cell*, 116 (2):281–297, 2004.

D. P. Bartel. MicroRNAs: target recognition and regulatory functions. *Cell*, 136(2): 215–233, Jan. 2009.

D. P. Bartel and C.-Z. Chen. Micromanagers of gene expression: the potentially widespread influence of metazoan microRNAs. *Nature reviews. Genetics*, 5(5): 396–400, May 2004.

A. Becskei, B. B. Kaufmann, and A. van Oudenaarden. Contributions of low molecule number and chromosomal positioning to stochastic gene expression. *Nature Genetics*, 37(9):937–944, Sept. 2005.

K. Bentele, P. Saffert, R. Rauscher, Z. Ignatova, and N. B. u. thgen. Efficient translation initiation dictates codon usage at gene start. *Molecular Systems Biology*, 9: 1–10, June 2013.

E. Berezikov. Evolution of microRNA diversity and regulation in animals. *Nature reviews. Genetics*, 12(12):846–860, Dec. 2011.

D. Betel, M. Wilson, A. Gabow, D. S. Marks, and C. Sander. The microRNA.org resource: targets and expression. *Nucleic Acids Research*, 36(Database issue): D149–53, Jan. 2008.

W. J. Blake, M. KAErn, C. R. Cantor, and J. J. Collins. Noise in eukaryotic gene expression. *Nature*, 422(6932):633–637, Apr. 2003.

J. Brennecke, A. Stark, R. B. Russell, and S. M. Cohen. Principles of microRNA-target recognition. *PLoS Biology*, 3(3):e85, Mar. 2005.

F. Buettner, K. N. Natarajan, F. P. Casale, V. Proserpio, A. Scialdone, F. J. Theis, S. A. Teichmann, J. C. Marioni, and O. Stegle. Computational analysis of cell-to-cell heterogeneity in single-cell RNA-sequencing data reveals hidden subpopulations of cells. *Nature Biotechnology*, Jan. 2015.

A. Castello, B. Fischer, K. Eichelbaum, R. Horos, B. M. Beckmann, C. Strein, N. E. Davey, D. T. Humphreys, T. Preiss, L. M. Steinmetz, J. Krijgsveld, and M. W. Hentze. Insights into RNA biology from an atlas of mammalian mRNA-binding proteins. *Cell*, 149(6):1393–1406, June 2012.

A. Chess, I. Simon, H. Cedar, and R. Axel. Allelic inactivation regulates olfactory receptor gene expression. *Cell*, 78(5):823–834, Sept. 1994.

L. Cong, F. A. Ran, D. Cox, S. Lin, R. Barretto, N. Habib, P. D. Hsu, X. Wu, W. Jiang, L. A. Marraffini, and F. Zhang. Multiplex genome engineering using CRISPR/Cas systems. *Science*, 339(6121):819–823, Feb. 2013.

V. T. Dang, K. S. Kassahn, A. E. Marcos, and M. A. Ragan. Identification of human haploinsufficient genes and their genomic proximity to segmental duplications. *European journal of human genetics : EJHG*, 16(11):1350–1357, Nov. 2008.

T. Davoli, A. W. Xu, K. E. Mengwasser, L. M. Sack, J. C. Yoon, P. J. Park, and S. J. Elledge. Cumulative haploinsufficiency and triplosensitivity drive aneuploidy patterns and shape the cancer genome. *Cell*, 155(4):948–962, Nov. 2013.

Q. Deng, D. Ramsköld, B. Reinius, and R. Sandberg. Single-cell RNA-seq reveals dynamic, random monoallelic gene expression in mammalian cells. *Science*, 343 (6167):193–196, Jan. 2014.

J. G. Doench, C. P. Petersen, and P. A. Sharp. siRNAs can function as miRNAs. *Genes & Development*, 17(4):438–442, Feb. 2003.

M. S. Ebert and P. A. Sharp. Roles for MicroRNAs in Conferring Robustness to Biological Processes. *Cell*, 149(3):515–524, Apr. 2012.

G. Edwalds-Gilbert, K. L. Veraldi, and C. Milcarek. Alternative poly (A) site selection in complex transcription units: means to an end? *Nucleic Acids Research*, 25(13): 2547–2561, 1997.

J. Elf and M. Ehrenberg. Fast evaluation of fluctuations in biochemical networks with the linear noise approximation. *Genome Research*, 13(11):2475–2484, Nov. 2003.

M. B. Elowitz, A. J. Levine, E. D. Siggia, and P. S. Swain. Stochastic gene expression in a single cell. *Science*, 297(5584):1183–1186, Aug. 2002.

A. J. Enright, B. John, U. Gaul, T. Tuschl, C. Sander, and D. S. Marks. MicroRNA targets in Drosophila. *Genome biology*, 5(1):R1, 2003.

J. T. Eppig, J. A. Blake, C. J. Bult, J. A. Kadin, J. E. Richardson, and Mouse Genome Database Group. The Mouse Genome Database (MGD): facilitating mouse as a model for human biology and disease. *Nucleic Acids Research*, 43(Database issue):D726–36, Jan. 2015.

L. Fagerberg, B. M. Hallström, P. Oksvold, C. Kampf, D. Djureinovic, J. Odeberg, M. Habuka, S. Tahmasebpoor, A. Danielsson, K. Edlund, A. Asplund, E. Sjöstedt, E. Lundberg, C. A.-K. Szigyarto, M. Skogs, J. O. Takanen, H. Berling, H. Tegel, J. Mulder, P. Nilsson, J. M. Schwenk, C. Lindskog, F. Danielsson, A. Mardinoglu, A. Sivertsson, K. von Feilitzen, M. Forsberg, M. Zwahlen, I. Olsson, S. Navani, M. Huss, J. Nielsen, F. Ponten, and M. Uhlén. Analysis of the human tissue-specific expression by genome-wide integration of transcriptomics and antibody-based proteomics. *Molecular & cellular proteomics : MCP*, 13(2):397–406, Feb. 2014.

K. K.-H. Farh, A. Grimson, C. Jan, B. P. Lewis, W. K. Johnston, L. P. Lim, C. B. Burge, and D. P. Bartel. The widespread impact of mammalian MicroRNAs on mRNA repression and evolution. *Science*, 310(5755):1817–1821, Dec. 2005.

H. B. Fraser, A. E. Hirsh, G. Giaever, J. Kumm, and M. B. Eisen. Noise minimization in eukaryotic gene expression. *PLoS Biology*, 2(6):e137, June 2004.

R. C. Friedman, K. K.-H. Farh, C. B. Burge, and D. P. Bartel. Most mammalian mRNAs are conserved targets of microRNAs. *Genome Research*, 19(1):92–105, Jan. 2009.

R. Fritsche-Guenther, F. Witzel, A. Sieber, R. Herr, N. Schmidt, S. Braun, T. Brummer, C. Sers, and N. Blüthgen. Strong negative feedback from Erk to Raf confers robustness to MAPK signalling. *Molecular Systems Biology*, 7:489, May 2011.

P. A. Futreal, L. Coin, M. Marshall, T. Down, T. Hubbard, R. Wooster, N. Rahman, and M. R. Stratton. A census of human cancer genes. *Nature reviews. Cancer*, 4 (3):177–183, Mar. 2004.

D. M. Garcia, D. Baek, C. Shin, G. W. Bell, A. Grimson, and D. P. Bartel. Weak seed-pairing stability and high target-site abundance decrease the proficiency of lsy-6 and other microRNAs. *Nature structural & molecular biology*, 18(10):1139–1146, Oct. 2011.

H. Goodarzi, H. S. Najafabadi, P. Oikonomou, T. M. Greco, L. Fish, R. Salavati, I. M. Cristea, and S. Tavazoie. Systematic discovery of structural elements governing stability of mammalian messenger RNAs. *Nature*, 485(7397):264–268, May 2012.

A. Grimson, M. Srivastava, B. Fahey, B. J. Woodcroft, H. R. Chiang, N. King, B. M. Degnan, D. S. Rokhsar, and D. P. Bartel. Early origins and evolution of microRNAs and Piwi-interacting RNAs in animals. *Nature*, 455(7217):1193–1197, Oct. 2008.

D. Grün, L. Kester, and A. van Oudenaarden. Validation of noise models for single-cell transcriptomics. *Nature Methods*, Apr. 2014.

H. Guo, N. T. Ingolia, J. S. Weissman, and D. P. Bartel. Mammalian microRNAs predominantly act to decrease target mRNA levels. *Nature*, 466(7308):835–840, Aug. 2010.

M. Hafner, M. Landthaler, L. Burger, M. Khorshid, J. Hausser, P. Berninger, A. Roth-baller, M. Ascano, A.-C. Jungkamp, M. Munschauer, A. Ulrich, G. S. Wardle, S. Dewell, M. Zavolan, and T. Tuschl. Transcriptome-wide identification of RNA-binding protein and microRNA target sites by PAR-CLIP. *Cell*, 141(1):129–141, Apr. 2010.

B. Haley and P. D. Zamore. Kinetic analysis of the RNAi enzyme complex. *Nature structural & molecular biology*, 11(7):599–606, July 2004.

D. Hanahan and R. A. Weinberg. The hallmarks of cancer. *Cell*, 100(1):57–70, Jan. 2000.

T. Hashimshony, F. Wagner, N. Sher, and I. Yanai. CEL-Seq: single-cell RNA-Seq by multiplexed linear amplification. *CellReports*, 2(3):666–673, Sept. 2012.

H. Herranz and S. M. Cohen. MicroRNAs and gene regulatory networks: managing the impact of noise in biological systems. *Genes & Development*, 24(13):1339–1344, 2010.

E. Hornstein and N. Shomron. Canalization of development by microRNAs. *Nature Genetics*, 38:S20–S24, June 2006.

G. Hutvágner and P. D. Zamore. A microRNA in a multiple-turnover RNAi enzyme complex. *Science*, 297(5589):2056–2060, Sept. 2002.

S. Islam, A. Zeisel, S. Joost, G. La Manno, P. Zajac, M. Kasper, P. Lönnerberg, and S. Linnarsson. Quantitative single-cell RNA-seq with unique molecular identifiers. *Nature Methods*, 11(2):163–166, Feb. 2014.

M. Jinek, K. Chylinski, I. Fonfara, M. Hauer, J. A. Doudna, and E. Charpentier. A programmable dual-RNA-guided DNA endonuclease in adaptive bacterial immunity. *Science*, 337(6096):816–821, Aug. 2012.

B. John, A. J. Enright, A. Aravin, T. Tuschl, C. Sander, and D. S. Marks. Human MicroRNA targets. *PLoS Biology*, 2(11):e363, Nov. 2004.

R. Kemkemer, S. Schrank, W. Vogel, H. Gruler, and D. Kaufmann. Increased noise as an effect of haploinsufficiency of the tumor-suppressor gene neurofibromatosis type 1 in vitro. *Proceedings of the National Academy of Sciences of the United States of America*, 99(21):13783–13788, Oct. 2002.

M. Khorshid, J. Hausser, M. Zavolan, and E. van Nimwegen. A biophysical miRNA-mRNA interaction model infers canonical and noncanonical targets. *Nature Methods*, Jan. 2013.

S. Klemm, S. Semrau, K. Wiebrands, D. Mooijman, D. A. Faddah, R. Jaenisch, and A. van Oudenaarden. Transcriptional profiling of cells sorted by RNA abundance. *Nature Methods*, 11(5):549–551, May 2014.

N. Kolesnikov, E. Hastings, M. Keays, O. Melnichuk, Y. A. Tang, E. Williams, M. Dylag, N. Kurbatova, M. Brandizi, T. Burdett, K. Megy, E. Pilicheva, G. Rustici, A. Tikhonov, H. Parkinson, R. Petryszak, U. Sarkans, and A. Brazma. ArrayExpress update–simplifying data submissions. *Nucleic Acids Research*, 43(Database issue):D1113–6, Jan. 2015.

A. Krek, D. Grün, M. N. Poy, R. Wolf, L. Rosenberg, E. J. Epstein, P. MacMenamin, I. da Piedade, K. C. Gunsalus, M. Stoffel, and N. Rajewsky. Combinatorial microRNA target predictions. *Nature Genetics*, 37(5):495–500, Apr. 2005.

E. C. Lai. Micro RNAs are complementary to 3' UTR sequence motifs that mediate negative post-transcriptional regulation. *Nature Genetics*, 30(4):363–364, Apr. 2002.

P. Landgraf, M. Rusu, R. Sheridan, A. Sewer, N. Iovino, A. Aravin, S. Pfeffer, A. Rice, A. O. Kamphorst, M. Landthaler, C. Lin, N. D. Socci, L. Hermida, V. Fulci, S. Chiaretti, R. Foà, J. Schliwka, U. Fuchs, A. Novosel, R.-U. Müller, B. Schermer, U. Bissels, J. Inman, Q. Phan, M. Chien, D. B. Weir, R. Choksi, G. De Vita, D. Frezzetti, H.-I. Trompeter, V. Hornung, G. Teng, G. Hartmann, M. Palkovits, R. Di Lauro, P. Wernet, G. Macino, C. E. Rogler, J. W. Nagle, J. Ju, F. N. Papavasiliou, T. Benzing, P. Lichter, W. Tam, M. J. Brownstein, A. Bosio, A. Borkhardt, J. J. Russo, C. Sander, M. Zavolan, and T. Tuschl. A mammalian microRNA expression atlas based on small RNA library sequencing. *Cell*, 129(7):1401–1414, June 2007.

B. Langmead and S. L. Salzberg. Fast gapped-read alignment with Bowtie 2. *Nature Methods*, 9(4):357–359, Apr. 2012.

R. Lee, R. Feinbaum, and V. Ambros. The C-Elegans Heterochronic Gene Lin-4 Encodes Small Rnas with Antisense Complementarity to Lin-14. *Cell*, 75(5):843–854, 1993.

Y. Lee, K. Jeon, J.-T. Lee, S. Kim, and V. N. Kim. MicroRNA maturation: stepwise processing and subcellular localization. *The EMBO Journal*, 21(17):4663–4670, Sept. 2002.

S. Legewie, D. Dienst, A. Wilde, H. Herzel, and I. M. Axmann. Small RNAs establish delays and temporal thresholds in gene expression. *Biophysical Journal*, 95(7): 3232–3238, Oct. 2008a.

S. Legewie, H. Herzel, H. V. Westerhoff, and N. Blüthgen. Recurrent design patterns in the feedback regulation of the mammalian signalling network. *Molecular Systems Biology*, 4:190, 2008b.

S. Legewie, C. Sers, and H. Herzel. Kinetic mechanisms for overexpression insensitivity and oncogene cooperation. *FEBS letters*, 583(1):93–96, Jan. 2009.

B. Lehner. Selection to minimise noise in living systems and its implications for the evolution of gene expression. *Molecular Systems Biology*, 4:170, 2008.

A. K. L. Leung, A. G. Young, A. Bhutkar, G. X. Zheng, A. D. Bosson, C. B. Nielsen, and P. A. Sharp. Genome-wide identification of Ago2 binding sites from mouse embryonic stem cells with and without mature microRNAs. *Nature structural & molecular biology*, 18(2):237–244, Feb. 2011.

E. Levine, E. Ben Jacob, and H. Levine. Target-specific and global effectors in gene regulation by MicroRNA. *Biophysical Journal*, 93(11):L52–4, Dec. 2007a.

E. Levine, Z. Zhang, T. Kuhlman, and T. Hwa. Quantitative characteristics of gene regulation by small RNA. *PLoS Biology*, 5(9):e229, Sept. 2007b.

B. P. Lewis, I.-h. Shih, M. W. Jones-Rhoades, D. P. Bartel, and C. B. Burge. Prediction of mammalian microRNA targets. *Cell*, 115(7):787–798, Dec. 2003.

B. P. Lewis, C. B. Burge, and D. P. Bartel. Conserved seed pairing, often flanked by adenosines, indicates that thousands of human genes are microRNA targets. *Cell*, 120(1):15–20, Jan. 2005.

Y. Li, Y. Sun, Y. Fu, M. Li, G. Huang, C. Zhang, J. Liang, S. Huang, G. Shen, S. Yuan, L. Chen, S. Chen, and A. Xu. Dynamic landscape of tandem 3' UTRs during zebrafish development. *Genome Research*, 22(10):1899–1906, Oct. 2012.

L. P. Lim, N. C. Lau, P. Garrett-Engele, A. Grimson, J. M. Schelter, J. Castle, D. P. Bartel, P. S. Linsley, and J. M. Johnson. Microarray analysis shows that some microRNAs downregulate large numbers of target mRNAs. *Nature*, 433(7027): 769–773, Feb. 2005.

P. S. Linsley, J. Schelter, J. Burchard, M. Kibukawa, M. M. Martin, S. R. Bartz, J. M. Johnson, J. M. Cummins, C. K. Raymond, H. Dai, N. Chau, M. Cleary, A. L. Jackson, M. Carleton, and L. Lim. Transcripts targeted by the microRNA-16 family cooperatively regulate cell cycle progression. *Molecular and Cellular Biology*, 27 (6):2240–2252, Mar. 2007.

J. A. Magee, S. A. Abdulkadir, and J. Milbrandt. Haploinsufficiency at the Nkx3.1 locus. A paradigm for stochastic, dosage-sensitive gene regulation during tumor initiation. *Cancer cell*, 3(3):273–283, Mar. 2003.

A. Marson, S. S. Levine, M. F. Cole, G. M. Frampton, T. Brambrink, S. Johnstone, M. G. Guenther, W. K. Johnston, M. Wernig, J. Newman, J. M. Calabrese, L. M. Dennis, T. L. Volkert, S. Gupta, J. Love, N. Hannett, P. A. Sharp, D. P. Bartel, R. Jaenisch, and R. A. Young. Connecting microRNA genes to the core transcriptional regulatory circuitry of embryonic stem cells. *Cell*, 134(3):521–533, Aug. 2008.

C. Mayr and D. P. Bartel. Widespread shortening of 3'UTRs by alternative cleavage and polyadenylation activates oncogenes in cancer cells. *Cell*, 138(4):673–684, Aug. 2009.

P. Mehta, S. Goyal, and N. S. Wingreen. A quantitative comparison of sRNA-based and protein-based gene regulation. *Molecular Systems Biology*, 4:221, 2008.

E. A. Miska, E. Alvarez-Saavedra, A. L. Abbott, N. C. Lau, A. B. Hellman, S. M. McGonagle, D. P. Bartel, V. R. Ambros, and H. R. Horvitz. Most Caenorhabditis elegans microRNAs are individually not essential for development or viability. *PLoS Genetics*, 3(12):e215, 2007.

S. Mukherji, M. S. Ebert, G. X. Y. Zheng, J. S. Tsang, P. A. Sharp, and A. van Oudenaarden. MicroRNAs can generate thresholds in target gene expression. *Nature Genetics*, 43(9):854–859, Sept. 2011.

J. R. S. Newman, S. Ghaemmaghami, J. Ihmels, D. K. Breslow, M. Noble, J. L. DeRisi, and J. S. Weissman. Single-cell proteomic analysis of S. cerevisiae reveals the architecture of biological noise. *Nature*, 441(7095):840–846, June 2006.

W. L. Noderer, R. J. Flockhart, A. Bhaduri, A. J. Diaz de Arce, J. Zhang, P. A. Khavari, and C. L. Wang. Quantitative analysis of mammalian translation initiation sites by FACS-seq. *Molecular Systems Biology*, 10(8):748, 2014.

J. Noorbakhsh, A. H. Lang, and P. Mehta. Intrinsic noise of microRNA-regulated genes and the ceRNA hypothesis. *PLoS ONE*, 8(8):e72676, 2013.

P. Oikonomou, H. Goodarzi, and S. Tavazoie. Systematic identification of regulatory elements in conserved 3' UTRs of human transcripts. *CellReports*, 7(1):281–292, Apr. 2014.

M. Osella, C. Bosia, D. Corá, and M. Caselle. The role of incoherent microRNA-mediated feedforward loops in noise buffering. *PLoS Computational Biology*, 7(3): e1001101, Mar. 2011.

E. M. Ozbudak, M. Thattai, I. Kurtser, A. D. Grossman, and A. van Oudenaarden. Regulation of noise in the expression of a single gene. *Nature Genetics*, 31(1): 69–73, May 2002.

J. Paulsson. Summing up the noise in gene networks. *Nature*, 427(6973):415–418, Jan. 2004.

J. M. Pedraza and J. Paulsson. Effects of molecular memory and bursting on fluctuations in gene expression. *Science*, 319(5861):339–343, Jan. 2008.

J. M. Pedraza and A. van Oudenaarden. Noise propagation in gene networks. *Science*, 307(5717):1965–1969, Mar. 2005.

K. S. Pollard, M. J. Hubisz, K. R. Rosenbloom, and A. Siepel. Detection of nonneutral substitution rates on mammalian phylogenies. *Genome Research*, 20(1):110–121, Jan. 2010.

A. Raj, C. S. Peskin, D. Tranchina, D. Y. Vargas, and S. Tyagi. Stochastic mRNA synthesis in mammalian cells. *PLoS Biology*, 4(10):e309, Oct. 2006.

N. Rajewsky. microRNA target predictions in animals. *Nature Genetics*, 38:S8–S13, 2006.

D. Ramsköld, E. T. Wang, C. B. Burge, and R. Sandberg. An abundance of ubiquitously expressed genes revealed by tissue transcriptome sequence data. *PLoS Computational Biology*, 5(12):e1000598, 2009.

J. M. Raser and E. K. O'Shea. Control of stochasticity in eukaryotic gene expression. *Science*, 304(5678):1811–1814, June 2004.

J. M. Raser and E. K. O'Shea. Noise in gene expression: origins, consequences, and control. *Science*, 309(5743):2010–2013, Sept. 2005.

A. Raue, C. Kreutz, T. Maiwald, J. Bachmann, M. Schilling, U. Klingmüller, and J. Timmer. Structural and practical identifiability analysis of partially observed dynamical models by exploiting the profile likelihood. *Bioinformatics*, 25(15):1923–1929, 2009.

B. J. Reinhart, F. J. Slack, M. Basson, A. E. Pasquinelli, J. C. Bettinger, A. E. Rougvie, H. R. Horvitz, and G. Ruvkun. The 21-nucleotide let-7 RNA regulates developmental timing in Caenorhabditis elegans. *Nature*, 403(6772):901–906, Feb. 2000.

I. L. Ross, C. M. Browne, and D. A. Hume. Transcription of individual genes in eukaryotic cells occurs randomly and infrequently. *Immunology and cell biology*, 72(2):177–185, Apr. 1994.

R. Sandberg, J. R. Neilson, A. Sarma, P. A. Sharp, and C. B. Burge. Proliferating cells express mRNAs with shortened 3' untranslated regions and fewer microRNA target sites. *Science*, 320(5883):1643–1647, June 2008.

J. M. Schmiedel, S. Klemm, Y. Zheng, A. Sahay, N. Blüthgen, D. S. Marks, and A. van Oudenaarden. MicroRNA control of protein expression noise. *Science (in press)*, Apr. 2015.

B. Schwanhäusser, D. Busse, N. Li, G. Dittmar, J. Schuchhardt, J. Wolf, W. Chen, and M. Selbach. Global quantification of mammalian gene expression control. *Nature*, 473(7347):337–342, May 2011.

M. Selbach, B. Schwanhäusser, N. Thierfelder, Z. Fang, R. Khanin, and N. Rajewsky. Widespread changes in protein synthesis induced by microRNAs. *Nature*, 455 (7209):58–63, Sept. 2008.

S. Serizawa, K. Miyamichi, and H. Sakano. One neuron-one receptor rule in the mouse olfactory system. *Trends in genetics : TIG*, 20(12):648–653, Dec. 2004.

P. Sood, A. Krek, M. Zavolan, G. Macino, and N. Rajewsky. Cell-type-specific signatures of microRNAs on target mRNA expression. *Proceedings of the National Academy of Sciences of the United States of America*, 103(8):2746–2751, Feb. 2006.

A. Stark, J. Brennecke, R. B. Russell, and S. M. Cohen. Identification of Drosophila MicroRNA targets. *PLoS Biology*, 1(3):E60, Dec. 2003.

A. Stark, J. Brennecke, N. Bushati, R. B. Russell, and S. M. Cohen. Animal MicroRNAs confer robustness to gene expression and have a significant impact on 3'UTR evolution. *Cell*, 123(6):1133–1146, Dec. 2005.

J. Stewart-Ornstein, J. S. Weissman, and H. El-Samad. Cellular Noise Regulons Underlie Fluctuations in Saccharomyces cerevisiae. *Molecular Cell*, 45(4):483–493, Feb. 2012.

A. O. Subtelny, S. W. Eichhorn, G. R. Chen, H. Sive, and D. P. Bartel. Poly(A)-tail profiling reveals an embryonic switch in translational control. *Nature*, 508(7494): 66–71, Apr. 2014.

P. S. Swain, M. B. Elowitz, and E. D. Siggia. Intrinsic and extrinsic contributions to stochasticity in gene expression. *Proceedings of the National Academy of Sciences of the United States of America*, 99(20):12795–12800, Oct. 2002.

Y. Taniguchi, P. J. Choi, G.-W. Li, H. Chen, M. Babu, J. Hearn, A. Emili, and X. S. Xie. Quantifying E. coli proteome and transcriptome with single-molecule sensitivity in single cells. *Science*, 329(5991):533–538, July 2010.

M. Thattai and A. van Oudenaarden. Intrinsic noise in gene regulatory networks. *Proceedings of the National Academy of Sciences of the United States of America*, 98(15):8614–8619, July 2001.

I. Ulitsky, A. Shkumatava, C. H. Jan, A. O. Subtelny, D. Koppstein, G. W. Bell, H. Sive, and D. P. Bartel. Extensive alternative polyadenylation during zebrafish development. *Genome Research*, 22(10):2054–2066, Oct. 2012.

M. C. Walters, S. Fiering, J. Eidemiller, W. Magis, M. Groudine, and D. I. Martin. Enhancers increase the probability but not the level of gene expression. *Proceedings of the National Academy of Sciences of the United States of America*, 92(15): 7125–7129, 1995.

L. M. Wee, C. F. Flores-Jasso, W. E. Salomon, and P. D. Zamore. Argonaute Divides Its RNA Guide into Domains with Distinct Functions and RNA-Binding Properties. *Cell*, 151(5):1055–1067, Nov. 2012.

E. K. White, T. Moore-Jarrett, and H. E. Ruley. PUM2, a novel murine puf protein, and its consensus RNA-binding site. *RNA*, 7(12):1855–1866, Dec. 2001.

J. K. White, A.-K. Gerdin, N. A. Karp, E. Ryder, M. Buljan, J. N. Bussell, J. Salisbury, S. Clare, N. J. Ingham, C. Podrini, R. Houghton, J. Estabel, J. R. Bottomley, D. G. Melvin, D. Sunter, N. C. Adams, Sanger Institute Mouse Genetics Project, D. Tannahill, D. W. Logan, D. G. MacArthur, J. Flint, V. B. Mahajan, S. H. Tsang, I. Smyth, F. M. Watt, W. C. Skarnes, G. Dougan, D. J. Adams, R. Ramirez-Solis, A. Bradley, and K. P. Steel. Genome-wide generation and systematic phenotyping of knockout mice reveals new roles for many genes. *Cell*, 154(2):452–464, July 2013.

B. Wightman, I. Ha, and G. Ruvkun. Posttranscriptional regulation of the heterochronic gene lin-14 by lin-4 mediates temporal pattern formation in C. elegans. *Cell*, 75(5):855–862, Dec. 1993.

Q.-L. Ying, M. Stavridis, D. Griffiths, M. Li, and A. Smith. Conversion of embryonic stem cells into neuroectodermal precursors in adherent monoculture. *Nature Biotechnology*, 21(2):183–186, Feb. 2003.

P. D. Zamore, J. R. Williamson, and R. Lehmann. The Pumilio protein binds RNA through a conserved domain that defines a new class of RNA-binding proteins. *RNA*, 3(12):1421–1433, Dec. 1997.

Acknowledgements

I feel fortunate that I could work under the supervision of three inspiring scientists. Their characters and their approaches to do and think about science could not be any more different, and working in these three (or rather four) different labs during my PhD was an invaluable experience.

I want to thank Nils Blüthgen for letting me join his group, great discussions, granting me a lot of freedom but always keeping me in check and supporting my multiyear endeavor to Boston. His ability to dismantle most of my 'great' ideas with just a few words inspired me to think harder but also saved me a lot of trouble.

I want to thank Debora Marks for introducing me to the world of microRNAs, her hospitality, our entertaining discussions about science and beyond and inspiring me to 'dumpster dive' through the inexhaustible wealth of published data.

I want to thank Alexander van Oudenaarden for giving me the opportunity to work in his lab when I had little to show for, teaching me to focus on the essentials and believing in my abilities.

I thank my colleagues in the ITB group for the wonderful and calm working atmosphere, especially Johannes, Kajetan, Pascal, Franzi, Bertram, Flo and Torsten.

I am in debt to my coworkers in the MIT and Hubrecht labs for their support and an inspiring scientific environment. Yannan for teaching me everything I needed to know about molecular biology and for contributing to my project; Sandy for great discussions about statistics and quantitative biology, late night support and our collaboration; Tim, Stefan, Magda, Nick, Nikolai, Dylan, Christoph, Mauro, Lennart, Abel, Kay and Corina for good times, good discussions and always helping me out when I approached them in panic, because something had not gone as not planned; and last but not least Dominic for insightful discussions about computational biology and fun times at Sterk.

I would also like to thank my master thesis supervisor Hanspeter Herzel for welcoming me into the ITB family and teaching me that sometimes 'knowing a bit of everything but nothing in particular' can actually be a good thing.

I am grateful for my friendship with Adrian, our discussions about science and life, our adventures, always providing me with a place I could call home, in Berlin as well as in Boston, and standing by me through hard times.

I thank my friends Christine, the Stefans, Anna, Sara, Vroni, Nadine, Katja, Torsten, Manu, Bine, Chong, Felix and Claudi for fun times, being there for me and putting up with my quirks.

Finally, I would like to thank my family for their love and support.